T0398480

DESIGNER SCIENCE

Designer Science

A History of Intelligent Design in America

C.W. Howell

NEW YORK UNIVERSITY PRESS
New York

NEW YORK UNIVERSITY PRESS
New York
www.nyupress.org

Library of Congress Cataloging-in-Publication Data
Names: Howell, C. W. (Christopher William), author.
Title: Designer science : a history of intelligent design in America / C.W. Howell.
Description: New York : New York University Press, [2025] |
Includes bibliographical references and index.
Identifiers: LCCN 2024040420 (print) | LCCN 2024040421 (ebook) |
ISBN 9781479827671 (hardback) | ISBN 9781479827701 (ebook) |
ISBN 9781479827718 (ebook other)
Subjects: LCSH: Intelligent design (Teleology)—History—21st century. |
Creationism—United States—History—21st century.
Classification: LCC BL262 .H68 2025 (print) | LCC BL262 (ebook) |
DDC 231.7/6520973—dc23/eng/20250113
LC record available at https://lccn.loc.gov/2024040420
LC ebook record available at https://lccn.loc.gov/2024040421

This book is printed on acid-free paper, and its binding materials are chosen for strength and durability. We strive to use environmentally responsible suppliers and materials to the greatest extent possible in publishing our books.

The manufacturer's authorized representative in the EU for product safety is Mare Nostrum Group B.V., Mauritskade 21D, 1091 GC Amsterdam, The Netherlands.
Email: gpsr@mare-nostrum.co.uk.

Manufactured in the United States of America

10 9 8 7 6 5 4 3 2 1

Also available as an ebook

To my parents, Scott and Karen Howell

What had that flower to do with being white,
The wayside blue and innocent heal-all?
What brought the kindred spider to that height,
Then steered the white moth thither in the night?
What but design of darkness to appall?—
If design govern in a thing so small.
—Robert Frost, "Design"

CONTENTS

LIST OF FIGURES

Introduction

Designer Science

One November day in 1996, John Angus Campbell boarded a van at Los Angeles International Airport. Campbell, a rhetoric and communications professor at Memphis University, was on the way to Biola University, where a large meeting of anti-Darwinists had commenced. It would be "an unprecedented intellectual event," wrote the decorated chemist Henry F. Schaefer III, "a major research conference bringing together scientists and scholars who reject naturalism as an adequate framework for doing science and who seek a common vision of creation united under the rubric of intelligent design."[1] Intelligent design (ID)—a broad, ideologically diverse, and theologically accommodating approach to anti-evolution—emerged onto the scene in the late 1980s and early 1990s. Hostility to evolution, and Darwin in particular, was of course not new, but ID represented an important departure from previous iterations, most notably young-Earth creationism. While riding in the shuttle, Campbell got a firsthand peek at the intellectual diversity within what was often called ID's "big tent," as well as the fragility of its alliances.

Religious opposition to evolution is usually described by the catch-all word "creationism." Even though it is a general term, it often has one meaning: young-Earth creationism (YEC). This is not the only kind of creationism there is, but slippage between the definitions is a frequent and exasperating problem in writing on the topic. Young-Earth creationism is typified by belief in a young Earth (six thousand years, give or take) and a literal approach to Genesis, both the Adam and Eve narrative (which is accepted as history, not mythology or allegory) and the global flood (which is presented as the explanation for the world's geological features). Yet, though it is often the loudest and most dominant form of creationism, it is far from alone. In the 19th and early 20th centuries, most creationists actually believed in a very old Earth, aged millions or possibly billions of years. In the second half of the twentieth century, the young-Earth version gained prominence, but it, too, had multiple incarnations. Biblical creationism (entirely based on the Bible) came first, and then scientific creationism (an attempt to build a scientific "model" with fewer overt biblical references) followed on its heels. Creationists attempted to get both versions into US schools; the courts defeated them each time. So, by the late '80s, a new alternative was needed. Intelligent design answered the call. Was intelligent design creationism? The answer is complicated. One can see its multifaceted nature up close by examining the disparate beliefs of those who got in the van with John Angus Campbell that morning in Los Angeles.

"First stop, who gets in but Paul Nelson," wrote Campbell, who kept a running diary of the conference. "Then Thane Ury (Bethel College) gets in. We start talking and then son-of-a-gun Paul says, 'There is Michael Denton'—I couldn't believe it." Denton, a British Australian biochemist and religious agnostic, was the author of the 1985 book *Evolution: A Theory in Crisis*, one of the first shots in ID's war against evolution. Campbell described him as a "lean 50-ish guy with a shock of white, close-cropped hair wearing a shirt that looks like the top for a pair of long underwear." Paul Nelson, a philosopher of science and the closest thing there is to a legacy creationist (his grandfather Byron C. Nelson, 1893–1972, was a well-known creationist writer from the early twentieth century), was the first sojourner to join Campbell in the van. "[Denton] and Nelson start dukeing [*sic*] it out right away," wrote Campbell. "It was fantastic." Though both Denton and Nelson supported ID, their ideological similarities ended there. Denton, it turned out, was something of an evolutionist, and he "proceeded to develop an evolutionary cosmology, the point of which is that there is abundant evidence for common descent [meaning that species today evolved from a common ancestor] and it is equally clear that evolution is directed and programmed."[2] Nelson seemed to retreat from the conversation, and Campbell speculated that he was "saving up for another time." Campbell concluded with the written equivalent of a shrug, observing, "Well, before I got to the hotel it was clear that 'intelligent design' is a general name for a diverse program."[3] ID could contain multitudes, from young-Earth creationists like Paul Nelson to non-Darwinian evolutionists like Michael Denton.

These disparate scholars and thinkers came together at Biola for what would later be called the "Birthplace of Design," a debut of sorts for intelligent design as its leaders sought to distinguish themselves from creationism and create an ideologically unifying strategy for unseating Darwin in contemporary science and culture. The timing was right, for 1996 had been a banner year for opposition to Darwin, perhaps surprisingly given how poor the outlook for anti-Darwinism had appeared the decade before. The latest incarnation of creationism—scientific creationism—had been roundly defeated in the 1987 *Edwards v. Aguillard* Supreme Court case. Scientific creationism was an attempt to make young-Earth creationism politically palatable to public schools. By emphasizing scientific models and softening biblical language, its proponents had hoped to get an airing in US high schools. They had failed. But, now, anti-evolution was back with a vengeance and with much greater scholarly bona fides. Intelligent design promised a new, alternative path forward, and its leading lights had been on the move.

In 1996 alone, the biochemist Michael Behe's book *Darwin's Black Box* was published by a major publisher and received attention in mainstream news outlets including *The New York Times*, the irascible mathematician and philosopher David Berlinski authored an explosively controversial (and explosively popular) essay for *Commentary* called "The Deniable Darwin," and a think tank in Seattle—the Discovery Institute—had opened a research center dedicated to intelligent design. Now, that same year, the Biola conference would spur the entire movement forward. With nearly two hundred attendees on location, its leaders set forth three goals: "unite on common ground," "build a community of thought," and "share ideas and knowledge."[4] The conference would be called "Mere Creation," and its goal was to achieve solidarity in the fractious and byzantine world of anti-evolutionism.

"If you're not overly scrupulous," wrote the intelligent design advocate David Klinghoffer, "you can take any idea and try to sell it to Christians by slapping the word 'Mere' on it." The adjective has an ecumenical and foundational connotation within Christian circles because of C. S. Lewis's *Mere Christianity*, and using the word for one's project "gives the impression that it would have been heartily endorsed by the most beloved of Christian apologists."[5] Klinghoffer was actually referring to the title of a book by Gregory Cootsona that was critical of intelligent design, but this only reinforces the point: the word "mere" is like a sweetener that makes the medicine go down.[6] It is a unifying, umbrella term, one capable of bringing together different denominations or people without any confessional entanglements. The 1996 "Mere Creation" conference was open to adherents of any faith or none, to skeptics of Darwin ranging from young-Earth creationists to theistic (and even some agnostic) evolutionists.

A shift in focus was evident. Rather than targeting science alone, or allowing the Bible to determine its scientific framework, ID's most accomplished proponents wanted to elevate the battle to philosophy and metaphysics. Their chief foe was not science; it was naturalism—the philosophical idea (and only philosophical, not scientific, they asserted) that nothing but nature exists. No supernature, no divinity. Intelligent design would resist this starting point. In defining the conference's mission, William Dembski—one of ID's most important thinkers, a scholar with dual PhDs in mathematics and philosophy, and an MDiv to boot—laid out four prongs: "a scientific and philosophical critique of naturalism," "a positive scientific research program, known as intelligent design," "a cultural movement for systematically rethinking every field of inquiry that has been infected by naturalism," and "a sustained theological investigation that connects the intelligence inferred by intelligent design with the God of Scripture and therewith formulates a coherent theology of nature." Dembski edited a volume titled *Mere Creation*, which was born out of the conference and featured the papers presented. He noted with optimism, "the possibilities for transforming the intellectual life of our culture are immense."[7]

This minimalist approach marked a shift from previous anti-evolutionary efforts, which had been wedded too strongly to the Bible to overcome the church-state divide and be accepted into secular classrooms. Even scientific creationism, which attempted to sequester the Bible, still allowed the sacred text to determine what was and was not scientifically licit. All of scientific creationism's "models" were arbitrated by Genesis. Critics, and the courts, did not fail to notice. ID, on the other hand, sought to transition from a biblically based anti-evolution to a more philosophical version—both rhetorically and ideologically. This shift was strategic as well as prudent, and it meant avoiding biblical arguments about Adam and Eve, the Noachian flood, and the age of the Earth. As Phillip Johnson (1940–2019)—University of California law professor and, for all intents and purposes, the "father" of intelligent design—concluded at the conference, "we also [must] put the biblical issues to one side," because "bringing the Bible anywhere near this issue just raises the *Inherit the Wind* stereotype and closes minds instead of opening them."[8] Johnson compared philosophical naturalism to a battleship. It might seem indefatigable and impregnable, "but the ship," he continued, "has sprung a metaphysical leak." If religious skeptics of Darwin could show a unified front, then perhaps the metaphysics of naturalism could be replaced by metaphysics hospitable to religious faith.[9] Darwin, Johnson believed, would not be fit to survive in such a competitive intellectual world, as evolution needed naturalism to support itself. Without naturalism, evolution would fall by the wayside, and Darwin's religious opponents could then hash out their biblical and theological differences from a secure position. However, such a lofty goal—the common alliance of all anti-evolutionists—would be hard to achieve, especially when some of them would be forced to shelve dearly held beliefs about the age of the Earth, the function of mutation, or the reality (or nonreality) of the common descent of every living organism from a shared ancestor, all in the name of expedience.

The last traveler in the van with Campbell, Denton, and Nelson was Thane E. Ury, a theologian and young-Earth creationist, who separately observed the same cracks in the alliance that Campbell saw. Overall, he was impressed not only by the speakers but also by the registrants, writing that the conference comprised "a veritable who's who from 'intelligent design theorists' worldwide." In fact, he noted, "one even quipped that the attendees were in some measure more impressive than the presenters." However, Ury was skeptical of the "big tent" mission of the conference. The approach was effective, he wrote, in allowing "colleagues to find common ground in rejecting methodological naturalism as a sound paradigm for science, while advocating a consensual objective of intelligent design as a major unifying theme for 21st century creationists." However, because young-Earth creationism was "given only inconsequential representation at best, overconfidence in solidarity may be only pyrrhic." Altogether, for many creationists, "the price of unity may be too high, since it has been bought at the cost of theological compromise on issues like the age of the Earth, prelapsarian suffering, death and extinction, and the hermeneutical perspicuity of Genesis."[10] If ID did not have a standard interpretation of Genesis, many creationists—like Ury—would find it unpalatable. Young-Earth creationists had long been adamant that death did not exist before Adam and Eve sinned in the Garden of Eden, but ID was agnostic on this issue. Could an alliance truly exist if they did not agree on matters of such bedrock import?

However, whereas for some young-Earthers ID was not creationist enough, for many critics—of both ID and creationism in general—it was too creationist already. At each stage in anti-evolutionary history, creationists of all stripes have attempted to get their ideas taught in US public schools—which is partly why they have tried to frame their ideas as purely scientific and not religious. In the efforts of ID's critics to refute intelligent design and prevent it from entering public curricula, most sought to prove that ID was nothing more than duplicitous creationism—"creationism in a cheap tuxedo," as one critic memorably put it, or "nothing more than standard-fare creationism gussied up to look like something it isn't."[11] In fact, when intelligent design ultimately found its way into the courtroom in 2005, in the *Kitzmiller v. Dover* court case, it was dealt a terrible blow when the federal judge John E. Jones III ruled that it was religion and not science primarily because of its association with creationism.[12] ID advocates long protested that they were not creationists, and the suspicion with which they were met by creationists themselves seemed to indicate a fundamental disagreement; but critics were less interested in the internal debates than they were with the threat ID posed to standard science. To defeat ID, showing that it was creationist was paramount.

Once successfully painted with that scarlet letter "C," intelligent design's star began to dim, and the years since the *Dover* court case were not kind to it. "In the nineties and early 2000s," wrote the mathematician Jason Rosenhouse, "ID seemed to be producing one novel argument after another. . . . It was [then] possible to wonder seriously if ID was a serious intellectual movement, or just another fad that would die out on its own. That verdict is now in. ID is dead."[13] Not to be outdone, David Klinghoffer asked whether the market for think pieces asking if "Intelligent Design is dead" was dead.[14] Whether or not ID is truly dead is an interesting question, but it is not as interesting as that famous litany of questions that every child in grammar school knows: What was it?

Where did it come from? What did it do? And, to adapt the legendary query posed by Drax the Destroyer, we shall do one better: Why was intelligent design?

While ID's community of supporters is still active—especially on the internet—in the media and culture at large, it is generally perceived today as defunct, and virtually all prior scholarly work on intelligent design ends with the *Dover* case from 2005. But intelligent design did not disappear. This book makes the case that its legacy is far more significant than we might think. Even though it did fail to achieve its goal of remaking contemporary science, intelligent design both planted the seeds and nurtured the growth of extreme skepticism in the world of US conservatism, a trend that has continued to grow ever since, sprouting in contemporary antivaccine movements and climate change denialism, among other things. We are all grappling today with intelligent design's legacy, even when we are not directly talking about its serious intellectual content and its highly technical scientific attacks on Darwin.

Design and Creationism: A Historiographical Overview

ID's relation to creationism is an unavoidable starting point to any discussion of the movement. The earliest full-length treatments of ID to come out of scholarly presses such as Oxford or MIT addressed this issue, but they came at it from a primarily polemical angle—the interest was mostly in refutation. The two most prominent examples of critical interactions were Robert Pennock's *Tower of Babel* (1999) and Barbara Forrest and Paul Gross's *Creationism's Trojan Horse* (first released in 2004 and updated in 2007 after the *Dover* trial). The historical research that these books documented was secondary to disproving ID's scientific claims; their primary goal was to reveal the scandalous legacy of scientific creationism at its heart. These works, though they included historical elements, evinced a rather activist take toward ID (with Forrest and Pennock even testifying against ID at *Dover*). On the other side, in Angus Menuge's short history of ID, he accused Forrest and Gross of conspiracy-mongering.[15] In 2003, the ID advocate Thomas Woodward penned a history of the movement that focused on the rhetorical and argumentative strategies of its leading figures: Michael Denton, Phillip Johnson, Michael Behe, and William Dembski.[16] The book is useful in that it collects a wide array of sources—Woodward is an ID insider and so had access to correspondence and interviews—but it also functions as an apologetic for ID and a hagiographic sketch of its leaders. Woodward's second book, *Darwin Strikes Back*, was even less impartial and served mostly as an extended critique of ID's opponents. The single most important event in the history of ID—the *Dover* case—was almost entirely ignored, aside from Woodward's criticism of Judge Jones III for his "egregious factual errors."[17]

While these full-length works on ID were polemical, the nonpolemical investigations into intelligent design have tended to include it as part of a story of creationism, rather than focusing on it separately. For instance, Edward Caudill, a journalism professor at the University of Tennessee, released a history of creationism in 2013 that included ID in its middle third. Caudill focused primarily on the court cases of *Scopes* and *Dover* and charted a historical trajectory of populism, media representation, and political campaigns against evolution, stretching from William Jennings Bryan (the presidential candidate and creationist champion at *Scopes*) to Phillip Johnson (the aforementioned "father" of ID). The various kinds of creationism and intelligent design, in his view, were similar enough to be seen as part of the same trajectory. ID had some differences but not enough to be chronicled on its own.[18] The standard overview of ID's history has remained a chapter in the second edition of *The Creationists* (2006), by Ronald Numbers's (1942–2023), which he added to the original work (published in 1992) after ID rose to greater prominence in the twenty-first century.[19] The upshot is that the major works on ID are either long refutations or short scholarly assessments. There has been no full-length history of ID.

This book is intended to fill this gap. It aims to sketch a historical overview of intelligent design while presenting a series of interlocking arguments. In so doing, it shows the trajectory of both ID's development and (supposed) demise and illuminates the ways in which we are still feeling its influence today. At its heart, ID was based on a radical idea: that one's metaphysics determines one's physics. This meant, then, that one's religious or nonreligious presuppositions and assumptions—about whether God exists, for example—had an inordinate and maybe even determinative effect on one's scientific ideas. If you believed in God, you might find intelligent design intuitively true, but if you did not believe in

God, then you would find yourself embracing Darwin. Evidence did not lead to theory; theory led to evidence. Consequently, embracing intelligent design also meant embracing deep doubt and even cynicism about the scientific establishment and its leading thinkers (how else to explain how they got it all so wrong?). Though ID was an intellectual movement that chose to combat Darwin on scientific grounds, it both fostered and intensified a culture of extreme skepticism about science and scientists. And so, when ID struggled to win support with its scientific arguments, what remained was this hyper-skeptical mindset. ID did not create it, but it did intensify it—and it has become its most lasting legacy.

Chapter 1 analyzes ID's relationship to creationism. Tracing ID's ideological origins shows that its relationship to creationism is like that of a sitcom romance: complicated. While it bears undeniable association with creationism, and argumentative overlap, the two cannot be easily considered the same thing, especially if creationism is specifically understood as young-Earth creationism. ID's differences from YEC are clear when held under a microscope, but ID *did* grow out of YEC, even if it rejected many of YEC's positions. After various species of creationism were defeated—including flood geology, which centered the Bible-based version of YEC, and later scientific creationism—ID rose up to take their places. However, while critics have alleged that ID was simply the same thing as scientific creationism but with a different name, these charges overlooked the true nature of ID's changes. ID grew out of the scientific revolutions in molecular biology and targeted well-known scientific problems such as the origin of the first cell. Furthermore, as ID emerged, it abandoned many of the principles that made creationism (at least in its most prominent forms) distinctive—most importantly the young Earth and the global flood from Genesis. ID demoted the Bible an arbiter of what scientific concepts were permissible to merely a side issue. One by one, ID embraced concepts anathema to young-Earth creationism: Big Bang cosmology, the old Earth, uniformitarianism (the idea, central to Darwin's thought, that observing nature in the present is the key to understanding the ancient past), and for some, even common descent. Some of these ideas are more acceptable to old-Earth creationists, like Hugh Ross—who spoke at the Mere Creation Conference. Ross, though, still emphasized the role of the Bible in science, and while he accepted the Big Bang, he rejected common descent (which some IDers affirm, like Behe or Denton). Typically, I will use the word "creationism" to mean the broad category of nonevolutionary worldviews and will use "young-Earth creationism" to specify that movement. Some confusion is unavoidable, as YEC is the most common type of creationism associated with the phrase and both critics and supporters use the terms rather interchangeably.

Instead of functioning merely as young-Earth creationism again, ID is more like a genus of anti-evolutionary thought, a broad framing like "mere creation" (or, in more common ID parlance, the "big tent") that can house almost any form of skepticism about evolution, but only through softening religious dogmas in favor of naturalistic critiques, critiques usable by old-Earth and young-Earth creationists and sometimes theistic or agnostic evolutionists (a few ID supporters accept evolution but doubt Darwin's version of it). In summation, both the critics and proponents of ID are right and wrong at the same time: ID is not simply "stealth" creationism, but it is not a fundamentally different movement either. This can easily lead to confusion. As Ronald Numbers observed in an interview, "I suspect most people don't understand what intelligent design is. I've talked to enough people who didn't understand it to make me really suspicious, and there are even people who lecture about it and don't know what they're talking about."[20]

What was ID then? Chapter 2 analyzes ID's philosophical, scientific, and theological components. Philosophically, ID was premised not on an explicitly biblical foundation but rather on a metaphysical presupposition—one of theism, the existence of God, and not naturalism. Phillip Johnson made the implicit philosophical assumptions of science his main target, and Darwinism, he alleged, was simply the result of naturalist metaphysics. Design was not permitted in science because, as Johnson saw it, the establishment watchdogs of scientific practice would never allow it a seat at the table. The issue was one of ontology, of a priori principles. How to demonstrate, then, that naturalist metaphysics was inadequate? Follow the playbook laid out by the philosopher of science Thomas Kuhn (1922–1996). A design theorist should highlight all the controversies in the current Darwinian paradigm, stressing it until it breaks down. This was the impetus behind ID's two most prominent scientific concepts: the biochemist Michael Behe's irreducible complexity and William Dembski's specified complexity (the most complicated and detailed ideas ID put forth, both of which argue—in different ways—that complexity found in nature implies a designing intelligence). Once Darwinism was demonstrated to be inadequate, then design would be able to compete as a replacement paradigm. However, apart from ID's metaphysics and science, most of its proponents were openly religious, so what was its theology? The old design argument, often associated with the natural theology of the eighteenth-century thinker William Paley, is the most natural starting point. Paley famously argued that, if one found a watch lying on the ground, one would logically and intuitively deduce it had been designed by a creator. And so, as with a watch, the complexity of any number of natural objects (such as the eye) also implied a creator. Paley then went further and sought to prove what this designer was like and who it was. As with creationism, ID's relation to Paley was likewise fraught with ambiguity. Michael Behe, for instance, warmly embraced both Paley and his design argument; however, William Dembski roundly denied that ID was a theological enterprise like Paley's was, even if a barebones version of the design argument might be useful. In examining this contrast, ID's strategic separation of reason and revelation—its attempt to sideline issues of religious dogma and biblical testimony in favor of science and philosophy—evinced a deeper strategic move. That is, ID was characterized at bottom by a hermetic seal between religion and science and between reason and revelation. These splits allowed ID to form broad alliances across all kinds of anti-evolution, embracing any and all skeptics by advocating only a baseline worldview that rejected naturalism. Such ideological minimalism enabled ID to create a broad, though shallow, political coalition.

Chapter 3 focuses on ID as politics. As a conservative movement, ID was characterized by its breadth, as it was able to draw support from across the right-wing spectrum, especially the intellectual strains. There were many conservatives in the twentieth century who were skeptical of Darwin but who did not want to embrace the seemingly anti-intellectual alternative of creationism, whether they were traditionalist figures such as Russell Kirk and Richard Weaver or neoconservatives such as Irving Kristol. All found Darwin unpalatable but for different reasons. A relatively sparse ideology like ID, framed only around anti-evolution and an unnamed intelligent designer, could and did garner favor with a wide array of conservative thinkers. The political momentum behind ID grew quickly in the 1990s, and by the early 2000s, it had influenced both cerebral intellectual conservatives and grassroots evangelical populists in their culture-war campaigns against US secularism, liberalism, and evolutionary theory. Most significantly, there was pervasive interest in changing public school curricula across the nation. Despite these campaigns, some ID leaders, especially William Dembski, worried that ID's political achievements would outrun the capability of its scientific program and could possibly damage the movement's vitality. This, it turned out, was exactly what happened in 2005 in Dover, Pennsylvania.

Chapter 4 covers not only the *Dover* case but also the backlash against ID that followed it. The court case was occasioned by a group of young-Earth creationists on the Dover Area School Board who worried about the influence that evolution was having on US culture and the religious and spiritual life of its children. They responded by trying to include intelligent design as an alternative in the school curriculum, a move that alarmed both the US scientific community and the ID leaders based out of the Discovery Institute think tank. While most ID proponents bowed out of the resulting courtroom battle, the case went on without them, and the result was a disastrous defeat for intelligent design. The hits kept coming after *Dover*, as the New Atheists (an energetic and zealous atheist movement that arose in the middle of the first decade of the twenty-first century) and theistic evolutionists (mostly those at the BioLogos Foundation, an organization dedicated to harmonizing evolution and Christianity) both began a counterattack against intelligent design. Furthermore, there was a substantial rejection of ID from conservative Catholics informed by Thomistic philosophy (the school of thought deriving from St. Thomas Aquinas, 1225–1274, which many Catholics interpret as being hospitable to evolutionary theory). While the debate between ID and Thomists was rather arcane, revolving around concepts of divine action and causation, it nevertheless highlights that even in the conservative Catholic world—a place one might expect to find sympathy for ID—there was at best a cool reception of its ideas.

Chapter 5 covers the life of ID after *Dover*. Since 2005, ID has hardened into a more inflexible ideology, and its leading thinkers and writers have doubled down on the same arguments they had made in the '90s, insisting that they were scientifically valid and that the only reason for their defeat was ideological prejudice. Even so, ID began to disappear from public discourse aside from a brief return in 2008 with the release of the pro-design documentary film *Expelled*. One might be forgiven, then, for believing that it was a temporary storm over US public discourse that has since passed. However, its legacy is more pronounced than most people realize, especially regarding its influence on the US public's understanding of science. Instead of restoring confidence in science among religious believers who rejected evolution, ID helped to undermine its core demographic's confidence, both in science as an enterprise and in scientists as reliable practitioners. Though ID purported to be a new, scientifically legitimate paradigm through which to investigate God's action in nature, it contained within it a fundamentally suspicious kernel, for at bottom, ID was based on a presuppositional approach to knowledge that meant that science was not ideologically neutral but was rather informed only by a priori philosophical principles that could not be established scientifically. ID, which had long campaigned against modernity's woolly ideologies, ended up striking a blow against modernity's last vestige of absolute truth: the authority and reliability of science. This attitude took form in a strong rejection of expertise—and not only scientific expertise. But this does not mean that ID became anti-intellectual. "Expertise" is a deceptively complicated word, and the kind of expertise against which ID proponents began to chafe was the kind typified by liberal technocracy and was rooted in ID's suspicion of bureaucrats, social engineers, and technicians, whom they see as forming the shadowy ruling class lurking behind Washington, DC, politicians. Because ID advocates had actual expertise themselves, their campaign against "The Establishment" carried some weight with conservative populists, who already doubted that the government knew what was best for them. However, even that is not the end of the story, for there is one area in which ID's critiques of established science and its political and financial trappings can find much wider cachet across political divides: their recent creation of the Walter Bradley Center for critical analysis of artificial intelligence.

John Angus Campbell's brief van ride to the airport—carrying skeptics of Darwin from various perspectives to one destination and then letting them out again—is a good encapsulation of what ID is, and it illuminates its strengths and weaknesses. For a few decades, a twinkling in the eye of history, ideologically incommensurate skeptics of Darwin briefly united and worked together under an umbrella movement. This "big tent," or "mere creation," was the chief ideological strategy of intelligent design. Within this big tent, one could find specific ID arguments based on complexity, divine design, and conservative political ideologies; one could find adherents who often did not agree with each other. It was a coalition, but as Campbell noted in his diary and Ury noted in his retrospective, it was a fragile one. It eventually did break down, and the consequences of ID's retreat from an ecumenical anti-evolutionary movement to just one of many highly skeptical attacks on contemporary science are still being felt.

A Personal Perspective

It is sometimes hard to avoid an obvious bias when writing, and quoting other writers, on an issue as loaded with pathos as creationism is, especially because both its proponents and critics so readily dispense with propriety. Ronald Numbers, at the end his introduction to *The Creationists*, expressed both consternation and amusement at the fact that "academics who would have no trouble empathetically studying fifteenth-century astrology . . . lose their nerve when they approach twentieth-century creationism." The "prevailing attitude" is often not dispassionate inquiry but rather a sense that "we've got to stop the bastards."[21] It cannot be gainsaid that ID proponents are equally fond of ad hominem, either. However, my goal is not to "stop the bastards" or to promote them; I am here neither to praise ID nor to bury it. But because it is the twenty-first century and Bulverism is our national pastime, I suppose it is necessary now to give a word about my "bias" before proceeding.

When the Greek poet George Seferis was asked what his worldview was, he quipped, "I'm sorry to say that I have no world view. . . . Perhaps you find that scandalous, sir, but may I ask you to tell me what Homer's world view is?"[22] I would like to affirm this same sentiment, but I suppose I cannot go that far. Nevertheless, a suspicion of all "isms" is probably the most fundamentalist intellectual trait I share, which is perhaps to be expected since I grew up in a young-Earth creationist setting in Southern California. Kent Hovind videos formed a critical part of my junior high science education. I was briefly interested in ID as a high school and college student before eventually migrating into the theistic evolution camp. I suspect that this trajectory will make everyone unhappy. I am currently a practicing Eastern Orthodox Christian; but I wear my dogmas as lightly as a hat, and I have no interest in promoting a particular viewpoint on creation, design, or theism. My views on the matter are more informed by Sergei Bulgakov (1871–1944) than by any of the figures discussed in these pages, and few would probably find him an authoritative source.[23] But I am writing as a historian, not a scientist, theologian, or philosopher, so any judgment on issues of that nature is beyond my ken. That all said, I have tried my best to present ID in a critical but fair light—not to disprove it (or prove it) but simply to understand it and its supporters. It is my hope that this book will, to appropriate Darwin's phrase about origins, "throw light" on a movement that exerted, and still does exert, great influence on life in the US and that it will help illuminate the beliefs, politics, and effects of ID such as they were and are in the late twentieth and early twenty-first century.

1

Creationism

Same Old Creationism?

The intelligent design journal *Origins & Design* received an angry letter to the editor in 1996: "Though you try to disguise your bible-based creationism with euphemisms like 'intelligent causation,' it is clear that your 'interdisciplinary, peer-reviewed quarterly' is just one more outlet for Christian apologetics and anti-evolution proselytization." The letter writer concluded by describing the journal as "the same old creationist bullshit dressed up in new clothes."[1] Creationist sympathizers of ID would sometimes take the same approach (though usually with less vulgarity) and collapse all anti-evolution into the young-Earth creationist mold. Ronald Numbers, at the end of his chapter on ID in *The Creationists*, described a chance encounter with a businessman on an airplane who was serendipitously reading Phillip Johnson's *Darwin on Trial*. When Numbers asked what he was reading, the traveler assured him "that Johnson (who never appealed to Noah's flood) undermined the case for evolution by showing how fossils originated in the biblical deluge."[2] For many people, both favorable to evolution and hostile to it, ID was just another bead on the string of creationism, albeit one with new polish. The differences between it and other forms were negligible. Swap flood geology for irreducible complexity.

ID's strongest academic critics likewise identified it as creationism redux, a pseudoscientific phoenix rising from the ashes of the 1980s court cases. Barbara Forrest and Paul Gross wrote that when "Biblical literalists" such as Duane Gish and Henry Morris were defeated in court, they were "eclipsed by a new brand of creationists who have absorbed a part of their following: the new boys are intelligent design promoters." In *Creationism's Trojan Horse*, favored adjectives for ID were "old" or "ancient," as Forrest and Gross emphasized its lack of novelty. ID was merely "a new variant of the old (anti)scientific creationism" or "simply a restatement of the ancient argument from design."[3] Victor Stenger wrote bluntly, "Intelligent Design is the new buzz word for what used to be called 'creation science.'"[4] Robert Pennock made a similar move in *Tower of Babel*, where he called ID "intelligent design creationism" and "new creationism."[5] It is easy to see how most of the distinctions between young-Earth creationism, old-Earth creationism, progressive creationism, and intelligent design can get blurred or even eliminated in favor of the blanket condemnation of all garden-variety creationism.

Forrest and Gross, who were typically more strident in tone than Pennock, sometimes labeled ID as "mere creationism" (in an homage to *Mere Christianity* and perhaps the "Mere Creation" conference)—a sort of general statement on anti-evolution. Even so, they also wrote that the "big tent" was "cozier than most people realize" and that ID was "much more conventional in its creationist views than most people realize."[6] Pennock, for his part, did try to draw boundaries between these categories, and he recognized that ID was broader than young-Earth creationism.[7] He constructed the model of a tower (of Babel) to distinguish between the different kinds of creationism—devoting time to the ideological and practical differences between young- and old-Earth varieties and how they related to intelligent design—but it was difficult to adhere to this distinction in practice. The geologist and priest (and anti-creationist) Michael Roberts attacked *Tower of Babel* in a review, taking issue with the way Pennock was "very loose" in his definition of creationism, allowing him to group too many different anti-evolutionists together too easily. "ID is not of the same ilk," Roberts wrote. He countered by defining ID as a spectrum, writing that it varied "from restrained YEC to something approaching theistic evolution."[8] The world of anti-evolution is complicated, and the labels are difficult to apply consistently to the different players; so it is perhaps unsurprising that such slippage exists. However, one important consequence lurked behind it all: critics argued, and usually persuaded others, that ID was nothing new, that it was the same old Wizard of Oz behind the curtain of new terminology.

ID advocates, for their own part, strenuously objected to the connection and tried to distance themselves from creationism for the same reasons that its opponents worked so hard to establish its linkage. To be a creationist was to be already defeated: scientifically, culturally, and, most important, legally. Michael Behe flat out rejected it, writing that he even accepted common descent, making him a theistic evolutionist (though not a Darwinian one).[9] William Dembski argued that ID was not related to creationism because it "nowhere attempts to identify the intelligent cause responsible for the design in nature" and that it had no "prior religious commitments."[10] Years later, in 2010, Dembski was criticized by the young-Earth creationist Tom Nettles for his book *The End of Christianity*, wherein Dembski affirmed that death predated Adam and Eve's sin, denied the global nature of the flood, and embraced an old Earth.[11] After Phillip Johnson's first ID book, *Darwin on Trial*, was brutally reviewed by Stephen Jay Gould, his newfound supporters at the Ad Hoc Origins Committee circulated a letter in his defense, arguing, "we think a critical re-evaluation of Darwinism is both necessary and possible without embracing young-earth creationism."[12] The critical tendency to reduce ID to nothing more than a new form of creationism (and the implication that this creationism was not substantially different from YEC) led to a somewhat frustrated essay by the philosopher and ID historian Angus Menuge, wherein he argued that, in light of the "clear contrasts between ID and traditional creationism, it seems plausible that the pejorative 'creationist' label is used chiefly to encourage an attitude of dismissive rejection, which avoids engagement with ID's proposals."[13]

Some ID opponents agreed with this assessment from time to time, as Eugenie Scott (the anthropologist and longtime director of the anticreationist organization the National Center for Science Education) did in a review of Johnson's *Darwin on Trial*, in which she began, "*Darwin on Trial* is an antievolution book, not a 'scientific creationism' book."[14] But this distinction was not always drawn; for instance, Barbara Forrest continued to argue that ID was only creationism manqué after the fateful 2005 *Dover* trial. In an essay called "Still Creationism After All These Years," Forrest contended, "Despite denials by proponents of intelligent design . . . that ID is creationism, critical analysis by scientists and scholars, as well as statements by the proponents of ID themselves, has established beyond any doubt ID's true identity as neo-creationism." Forrest repeatedly scorned ID for its creationist heritage and condemned its proponents for engaging in "disingenuous stock quotations prepared for mainstream media." Despite the strong language, Forrest was not always clear in defining ID and alternated between calling ID creationism *tout court* and calling it a "direct lineal descendant" or referring to its direct ancestry, depending on the context.[15] But evolution, more than anything else, should teach us that there can be a large difference between a thing and its "direct ancestor." Related, yes—but that is the whole question: How so? Relation does not entail identity. Humans and the other apes are directly related and share a common ancestor, but evolution shows that, over time, minuscule variations can become big differences.

That ID might have strayed far from its creationist roots was evident in the way a great many young-Earth creationists rejected it and how they sought to either minimize its influence or reinforce the boundaries between them. In a four-part viewpoint series published by Zondervan, the arch-creationist Ken Ham, president of the YEC organization Answers in Genesis, put the matter bluntly: "contrary to what many secularists think, [the theory of intelligent design] is not simply repackaged biblical creation."[16] Ham distinguished between ID arguments—those about, say, the complexity of organic life that might indicate a divine designer, which creationists would accept—and the ID movement. The latter, for him, was no help at all to the cause. In Ham's view, ID was nothing more than the barest minimum rejection of naturalism, for ID could countenance both an ancient Earth and even theistic evolution: "only *atheistic* evolution is rejected." Ham did not write this in order to present a unified face in front of critical media, either; he responded directly to the ID luminary and philosopher Stephen Meyer, lambasting him and the rest of the ID movement for leaving out the Bible. "This strategy," he contended, "of arguing for design but ignoring Genesis was tried in the early nineteenth century but it failed to derail the deism and atheism that were increasingly taking over science." When Meyer extolled ID's success by pointing to the conversion of the legendary atheist philosopher Antony Flew (1923–2010) to deism, mostly because of ID's arguments, Ham responded that this meant nothing for the salvation of Flew's soul. "Having rejected or never considered the overwhelming evidence that the Bible is God's Word," wrote Ham, "[Flew] died, apparently, as a Christ-rejecting sinner who sadly will spend eternity in hell."[17] The content of one's religious beliefs matters in the grand soteriological scheme, and so many ardent young-Earth creationists found no succor in ID. Henry M. Morris (1918–2006), one of the most influential creationists of the twentieth century and the founder of the Institute for Creation Research, wrote, "Getting people to believe in 'intelligent design' is, therefore, neither new nor sufficient. People of almost every religion (except atheism) already believe in it."[18] The old-Earth creationist philosopher Norman Geisler (1932–2019) criticized the ambiguity of Behe's second book, *The Edge of Evolution*, and wrote that he could not accept many of its proposals because "Behe makes it clear that he is a theistic evolutionist."[19] Even the young-Earth creationists *within* the ID camp—such as Paul Nelson or John Mark Reynolds—stood their ground against their fellow ID colleagues when internecine disputes arose.[20] The differences between ID and other forms of creationism might have been negligible to their critics, but let no one say that they did not matter to those who were on the inside.

Where does the truth lie in all of this, then? ID's critics were certain that intelligent design was simply creationism with a mask, while creationist critics wondered why ID proponents did not admit that they were evolutionists and cease lying to themselves and others. Who was right? And furthermore, why was it so crucial for ID's critics to establish that it was "stealth creationism," even when many creationists themselves were eager to disassociate? The importance of the legal system cannot be overlooked. The historiography of creationism is dominated by court cases. From *Tennessee v. Scopes* (1925) to *Epperson v. Arkansas* (1968) to *Edwards v. Aguillard* (1987), the courtroom has been the battleground of choice for this conflict. The cases of the late '60s to the late '80s set precedent, denying both efforts to keep evolution from being taught in public schools (*Epperson*) and to teach creationism along with it (*Edwards*) so that linking any other form of anti-evolution to creationism would result in an immediate judicial victory. Showing that ID was the "same old creationist bullshit" or fundamentalism run amok was the most expedient way to eradicate it from the US body politic.

Is, then, ID the same thing as creationism? The answer is a resounding: that depends on what is meant by "creationism." ID is certainly creationist in a broad sense, but in the restricted young-Earth variety far less so. There is a great degree of slippage regarding the term "creationist," and it has been claimed by an array of anti-evolutionary thinkers who often had conflicting religious and scientific commitments. It has also been imputed to critics of Darwin who did not view themselves as creationists, or even religious, at all. "Formerly," writes Christopher Rios, "the term *creationist* carried a variety of meanings ranging from those who believed that each human soul was specially created to those who opposed evolution or accepted special creation of species." In the mid-twentieth century, things changed. "Only after George McCready Price began promoting his flood geology," Rios continues, "did the term come to be reserved for those who insisted on young-earth creationism."[21] Yet it is very easy to overlook this. Ronald Numbers notes that even he, "in a 450-page book on the history of modern creationism, failed to address the issue of when these terms first came into use." In point of fact, "until well into the twentieth century [Americans] rarely used the term creationism," and when they did, the meaning was so ambiguous that "it is often difficult to distinguish an evolutionist from an antievolutionist."[22] If these terms are difficult for historians to parse, just as much struggle can be expected by the players on the field, and so it should not surprise us that opponents and supporters of evolution often neglected to define what they meant with the word "creationist." Darwin inveighed several times against the "ordinary view of independent creation," in *The Origin of Species*, but as Numbers observes, at the time Darwin wrote, this phrase often meant a rejection of traducianism (that an individual's soul is passed on and derived from one's parents) and was not always something compared to or contrasted with evolution in biology.[23] As Rios mentions, it was not until nearly a century after the *Origin* that flood geology, the emergent form of young-Earth creationism that focused on the biblical record, effectively monopolized the use of the term "creationist." Even that form, however, eventually shifted into something new, with the advent of "scientific creationism" in the 1970s and '80s (and, with it, an infusion of ideas from molecular biology and biochemistry).

Ultimately, as we will see, ID contained some elements of each era of creationism, with clear similarities and differences between it and early fundamentalist anti-evolution, flood geology, and scientific creationism. There was notable philosophical and scientific overlap with the beliefs of early American fundamentalists, most obviously in an affirmation of the old Earth, an epistemological embrace of Common Sense Realism, and an antipathy to professional expertise and credentialism. The mid-twentieth-century flood geology of John C. Whitcomb and Henry M. Morris (famously championed today by Ken Ham) is the most removed from ID, for design advocates largely do not share its concerns with biblical literalism, the Fall of Man, or the global flood (concepts that are of critical importance to YEC but that are nevertheless either sidelined or outright rejected by most IDers). However, while ID similarly differed from later "scientific creationism" on the age of the Earth, some of ID's most famous arguments—the biochemical argument against abiogenesis (the origin of the first cell from nonliving matter) and the concept of specified complexity—were nevertheless incubated by the scientific creation movement and were nourished to maturity as creationists considered the development of biochemistry and chemical evolution research in the twentieth century.

When put under a microscope, ID can be seen as an eclectic anti-evolutionary movement that is based primarily on minimalist theological or doctrinal claims, an interest in more old-Earth-friendly scientific ideas (especially the Big Bang theory), and a renewed emphasis on the argument from design. It was nurtured during the heyday of scientific creationism and the legal contests over creationism, before ultimately emerging from its chrysalis in the 1990s as a distinct form of anti-evolution. As time progressed, ID shed more and more of its creationist trappings—abandoning the distinctive features that made creationism, especially young-Earth creationism, what it was. At its root, ID is still religiously informed anti-evolution, but its expression and ideology developed in new ways.

Fundamentalism and Anti-evolution

The word "fundamentalist" is, like "creationist," fraught in meaning and has changed over time. Recently, the *Dune* movies started calling the Fremen "fundamentalists," evidence that the word has become fully divorced from any historical meaning or context. And the situation in academia is often not that different from in pop culture. "On the most contemporary academic use of the term," writes the Christian philosopher Alvin Plantinga, "it is a term of abuse or disapprobation, rather like 'son of a bitch.' . . . When the term is used in this way, no definition of it is ordinarily given."[24] But there is a historical source for the word and a movement from which it arose.

In the early twentieth century, American fundamentalism emerged to protect what its proponents took to be the core doctrines of Christianity, and there were two opposing issues that bedeviled them more than any others: biblical higher criticism and evolution. These perceived forces of "modernism" led them to internal battle with their coreligionists in what is now known as the Fundamentalist-Modernist Controversy. The most iconic event of the period was the 1925 *Scopes* "Monkey trial" in Dayton, Tennessee. John T. Scopes (1900–1970), a football coach and teacher, was put on trial for teaching evolution in his public school, a violation of Tennessee's Butler Act. The resulting trial was a circus (the celebrity lawyer Clarence Darrow, 1857–1938, defended Scopes, while the anti-evolutionist banner was taken up by William Jennings Bryan, 1860–1925, former secretary of state and perennial presidential candidate). In the aftermath, the fundamentalist cause was badly damaged. However, as George Marsden chronicles in *Fundamentalism and American Culture*, the fundamentalists actually won at the *Scopes* trial (Scopes was guilty, after all). But they lost when it came to the battle of American hearts and minds. Marsden characterizes this as a kind of Kuhnian paradigm shift, one in which the old, fact-gathering and inductive method of Baconian science gave way to a new form of (rather Kantian) interpretative and speculative hypothetical science.[25] But this was about more than just science. Common Sense Realism, a philosophically realist stance in which everyone has innate ability to understand issues of science and biblical interpretation, was integral to the Presbyterian institutions that were increasingly riven by the fundamentalist-modernist split. It influenced not only the way they saw nature but also the way they read the Bible. Central to Protestant intellectual structure is *sola scriptura* and the importance of individual interpretation of the sacred text, ungoverned by any hegemonic interpretative system like the Catholic magisterium. It must be able to be read as self-evident and clear truth—just as nature should be. As Marsden writes, "When Scripture was looked upon as the compellingly perfect design of God, every detail was significant."[26] Design, then, of both nature and of scripture was central to this intellectual edifice, and so one frequently finds it in the anti-Catholic exegesis of fundamentalists. Writes Marsden, "the idea that a person of simple common sense could rightly understand Scripture was grounded in the more general affirmation of the Scottish philosophy that in essentials the common sense of mankind could be relied upon."[27] One also finds this idea in Thomas Reid (1710–1796), the Scottish exponent of Common Sense Realism who advocated a design argument for God's existence.[28] Fundamentalist anti-evolutionists hewed closely to common sense in both arenas, and in general, this meant an antagonism toward what they termed "science-falsely-so-called," the kind of science driven not by common sense and obvious interpretation of basic facts but by nefarious "hypotheses" or, as William Jennings Bryan called them, "guesses strung together."[29]

Hostility to hypothetical science was part and parcel of a dominant sentiment of anti-expertise in the world of anti-evolution. According to Marsden, evolutionary theory was perceived as anti-democratic and elitist by fundamentalists. J. Gresham Machen complained, for instance, that experts had too much influence on US schools and culture.[30] It was both funny and depressing to fundamentalists that PhDs—who had so much learning—would so easily forget the commonsense obviousness of nature and the Bible.[31] The historian Edward B. Davis highlights this trend, too, quoting James R. Moore: "[Creationism was] at least in part a reaction against the triumphal positivism and overweening professionalism of established scientific authorities."[32] Davis highlights the 1930 debate between the creationist Harry Rimmer (1890–1952) and the chemist Samuel Schmucker (1860–1943), in which the audience seemed split between two different conceptual worlds, each holding to different standards of evidence and decrying the other as obtuse. Rimmer attacked the scientific expert as "a man who knows more and more about less and less" and "a man who knows everything about nothing."[33] For Davis, neither were the evolutionists wholly innocent of such posturing. In fact, Davis argues that both fundamentalists and their scientific opponents adhered to a kind of "folk science," more of an epistemological standpoint about evidence, authority, and meaning.[34] Its primary purpose is comfort and reassurance in the face of a confusing and mysterious world. Writes Davis, "To a significant degree, professional scientists have cultivated an image of polished, unchallenged authority that empowers them at the expense of others. This image, integral to academic folk science, is projected by science education in public schools, leading to resentment on the part of fundamentalists and many others who see themselves as lacking a voice in determining how their children are to be indoctrinated."[35] This ideological and emotional epistemology would continue to manifest itself in the history of anti-evolution, from the fundamentalists down to the intelligent design advocates of the present (and their New Atheist opponents, too). The difference, as we will see later, is that ID proponents were themselves also highly credentialed, and so their aversion to expertise was more that of the jilted insider than the jealous outsider. They were often most suspicious of people in their own guilds. Their legitimization of suspicion, however, would find widespread acceptance among more populist opponents of Darwin.

While ID would deploy the plea for "common sense" at great length later—and share much of the negative attitude toward expertise that the fundamentalists had—they would also depart from early anti-evolution. ID had little interest in the Baconian inductive science that characterized fundamentalist thought and embraced hypothetical, theory-driven science, albeit one that was based on different methodological (and implicit metaphysical) principles than mainstream science. That said, some fundamentalists were more open to evolution (provided that Darwin's seeming randomness was softened). Some of *The Fundamentals* pamphlets (from which fundamentalists got their name) were even penned by theologians open to evolution. As Norman Geisler admitted, biblical inerrantists and fundamentalist intellectuals such as B. B. Warfield, A. A. Hodge, and James Orr "embraced theistic evolution."[36]

However, the term "fundamentalist"—and "creationist"—would later become constricted to mean young-Earth creationism. This form of anti-evolution would form a kind of interlude between the more design-centric forms of early anti-evolution and later intelligent design. But while young-Earth creationism's theory of the universe was much different than ID's, it would also lay the groundwork for some of ID's most important ideas.

Out with the Old, In with the Young

When William Jennings Bryan took the stand that hot summer afternoon of 1925, in Dayton, Tennessee, he enabled Clarence Darrow to snatch victory from the jaws of defeat. John T. Scopes was found guilty of teaching evolution and fined, and the hopes of the nascent American Civil Liberties Union (ACLU) to take the case to the Supreme Court were dashed when the Tennessee legislature threw out the verdict based on a technicality (the judge had levied the fine rather than the jury). Legal result aside, the social fallout of the case was more important than the ruling. As the historian Edward J. Larson documented in his award-winning *Summer for the Gods*, the long-term consequences of the case seriously damaged both the anti-evolution movement and those who were defending Darwin. Initially, it was Darrow who suffered, since he came across as something of a bully, but as the years went by, it was Bryan whose image was most tarnished. By then, the book *Only Yesterday* and the play *Inherit the Wind* had solidified the popular view of anti-evolutionists as backward country bumpkins whose ignorance of science was only matched by their ignorance of history.[37] However, despite the acerbic mockery poured on Bryan by H. L. Mencken (1880–1956), his interrogation by Darrow, and the circus-like nature of the trial, the huge public backlash against evolutionary theory persuaded most textbook makers and evolutionarily minded schoolteachers to avoid wading too deeply into the topic. Best to leave it for the colleges and preserve a fragile peace in the rapidly growing world of US high school. As a result, evolutionary biology remained largely untaught in schools until the late 1950s. What changed? There was a growing awareness of the poor state of US science education, there was some anxiety over competition with the Soviet Union in the Cold War, and there was the evolutionary modern synthesis, which knit together Darwinian evolution with Mendelian genetics.[38] Exasperated by the situation in schools, the geneticist H. J. Muller (1890–1967) exclaimed near the midcentury mark, "One hundred years without Darwinism are enough."[39] He would not have to wait much longer, for during the 1959 centennial of *The Origin of Species*, the slow integration of Darwinism and Mendelian genetics made its public debut.[40] It would be the end of Darwin's absence from US curricula, but it would be the beginning of a new phase in the battle. The uneasy calm between 1925 and the 1959 centennial would be broken by a new kind of anti-evolution: young-Earth creationism.

The flood geology that was promoted by young-Earth creationists was a radical departure from former styles of anti-evolution. The old-Earth variant had been the dominant perspective until the midcentury—it was the form that Bryan had adopted, as well as other fundamentalists like John Roach Straton (1875–1929) and Harry Rimmer. Bryan held, rather than an unflaggingly literal interpretation of Genesis, to the "Day-Age" theory, which was popular in the late nineteenth century as an interpretation of Genesis that fit with the uniformitarian evidence indicating an old Earth. What most concerned Bryan about evolution was its moral consequences, especially if humanity evolved.[41] While this sounds today like the desperate pleadings of an entrenched opponent, such a position put Bryan on hallowed ground with some scientists of his era; even Alfred Russel Wallace (1823–1913), co-discoverer of natural selection, rejected the applicability of natural selection to humanity's intellectual and moral character.[42] Bryan also equivocated on whether the Bible implied a young Earth. During the *Scopes* trial, he defaulted to illustrative interpretations of Genesis when Clarence Darrow pressed him. When asked whether he believed the Earth was made in six twenty-four-hour days, Bryan responded, "Not six days of twenty-four hours." He later elaborated, "I think it would be just as easy for the kind of God we believe in to make the earth in six days as in six years or in 6,000,000 years or in 600,000,000 years. I do not think it important whether we believe one or the other."[43] Nor was Bryan sui generis. The *Scofield Reference Bible* (the ultimate guide for fundamentalist readers of the Bible) drew attention in its notes to the "gap theory" interpretation of Genesis 1, which attempted to harmonize Genesis with an old Earth.[44] Lastly, the political motivations behind Bryan's crusade against evolution were hardly what one would consider today to be conservative. Long a progressive politician and advocate of the "common people," Bryan worried about the consequences that unrestrained ideas about the "survival of the fittest" would have for the general population of American workers (he was alerted to the influence of social Darwinism on German militarism by *Headquarters Nights*, by the entomologist Vernon Kellogg, 1867–1937). Bryan was a pacifist and anti-imperialist, and the rise in social Darwinist thought in the US profoundly unnerved him.[45]

One avenue of Bryan's attack was that scientists were increasingly doubting the validity of evolution by natural selection. With the benefit of hindsight, this seems to misrepresent the direction science was going, but at the time, there was more substance to Bryan's claim. Darwin did not know by what mechanism heredity was effected. He managed to convince the world of the validity of evolution and common descent, but he had a much harder time persuading many (even those who advocated his theories most forcefully, like Thomas Henry Huxley, 1825–1895, who privately was less enthusiastic than he was in public) of the centrality of natural selection, his chief idea.[46] Even until the late nineteenth and early twentieth centuries, many of the world's foremost scientists held to neo-Lamarckian views of the inheritance of acquired characteristics. Julian Huxley (1887–1975) called this period "the eclipse of Darwinism."[47] Genetics—the elusive "missing science of heredity"—did not fully emerge until rediscovery in 1900 of the pea-plant experiments of Gregor Mendel (1822–1884), and even then, Mendelianism was just as often seen as corrosive to Darwinian thought as it was supportive. William Bateson (1861–1926), who popularized the word "gene" and took Mendel's ideas public, challenged the gradualist view of Darwinism. Most Darwinians were gradualists—that is, they held the view that slow, incremental variations would produce, over time, the extraordinary variation of species that we now see. Some later interpreters were "saltationists" who believed that evolution took place via (comparatively) rapid "jumps." Bateson was of this camp.[48] Bateson's criticisms of Darwin were strong enough, and public enough, that Bryan noticed them and cited Bateson as an authority on the dubious truth of Darwinism, summarizing from him that "science has faith in evolution but doubts as to the origin of species."[49]

However, in the post-*Scopes* era, in the quiescent period of US education when evolution often went unmentioned, a revolution in biology was taking place. The emerging science of biochemistry would lead to renewed study of chemical evolution, and the new science of genetics would flower into the mathematical theory of population genetics, formalized by J. B. S. Haldane (1892–1964), Ronald A. Fisher (1890–1962), and Sewall Wright (1889–1988). With the edifice solidified and the framework built, Theodosius Dobzhansky (1900–1975), Ernst Mayr (1904–2005), and G. Ledyard Stebbins (1906–2000) led the charge and assembled the synthetic theory of evolution: the modern synthesis—genetics and evolution harmonized, Darwin and Mendel made allies and bedfellows at last. Science's long nightmare was over, and evolution had been rescued from its non-Darwinian interlude. As Julian Huxley wrote in his 1942 magnum opus, *Evolution: The Modern Synthesis*, "The death of Darwinism has been proclaimed not only from the pulpit, but from the biological laboratory; but, as in the case of Mark Twain, the reports seem to have been greatly exaggerated, since to-day Darwinism is very much alive."[50]

While evolutionary theory was itself evolving, creationism would mutate along with it. In the years after *Scopes*, the creationist and hydraulics engineer Henry Morris grew frustrated by the tendency of Christian and science organizations to capitulate one by one to either an old Earth or theistic evolution. Like the endless temptation toward polytheism in the Old Testament, American Christians continually reverted to ideas that Morris found unacceptable. Most alarming was when the American Scientific Affiliation fell to theistic evolution in the midcentury.[51] He was likewise frightened by the 1959 Darwin Centennial Celebration, writing later, "Creationism, except for isolated pockets of fundamentalists, seemed dead—most certainly among scientists! The immense and favorable publicity accorded to the Darwinian Centennial year, especially the great Darwin 'worship service' at Chicago, where speaker after speaker rhapsodized about Darwin's contribution to the life of mankind, and exhorted each other and all their disciples on to further glories of evolutionary achievement, seemed to be the final nail in the coffin of creationism and even of meaningful Christian theism."[52] Morris, then, sought to lead his people out of nearly forty years of wandering and shepherd them into the safe haven of "Biblically sound" creationism.[53] Rising in defense of "meaningful Christian theism," Morris partnered as junior author with the Old Testament scholar and theologian John C. Whitcomb (1924–2020). Together, they produced the seminal work of young-Earth creationism. *The Genesis Flood* was published in 1961 and effectively constricted the meaning of "creationist" from broad anti-evolution to an explicitly young-Earth worldview that focused on flood geology. Finding its origin in the thought of the Seventh-day Adventist George McCready Price (1870–1963), flood geology was taken to the fundamentalist mainstream by Whitcomb and Morris (despite their debt to Price, the persistent Adventist influence on young-Earth creationism would remain mostly unacknowledged by YECs in the century hence). As Numbers wrote, by their partnering together, "thus was sealed the pact that would soon make 'Whitcomb and Morris' a byword among evangelical Christians."[54]

One of the most striking things about *The Genesis Flood* is how little both Charles Darwin and evolutionary theory factored into it. Darwin did not even get mentioned by name until a footnote almost one hundred pages into the book. Instead of targeting Darwin or evolutionary biology directly, Whitcomb and Morris based most of their discussion on geology, the historicity of the Noachian deluge, and the role of Genesis as a scientific text. The purpose of *The Genesis Flood* was to restore catastrophism and the global flood to the center of Christian interpretations of both science and the Bible. Much of the first half of the book, surely disorienting to any scientist who was to pick it up for the purposes of refutation, was devoted to a discussion of the biblical record of the flood, descending into obscure niches of fundamentalist thought. Whitcomb and Morris took special care to reject the "local flood" theory, the idea that the flood described in Genesis only affected a particular and relatively small area in the ancient Near East (a view that, as noted earlier, some ID advocates would embrace). After laying out all the evidence against the local flood on a scriptural level, they argued that the burden of proof must lie with the local flood advocates, because the Bible's clarity on the global nature of the flood was so stark and incontrovertible that "the burden of proof" rested on those who doubted it. In fact, because of the nature of the Bible's testimony and its status as an authoritative word from God, they argued, "we do believe that no problem, be it scientific or philosophical, can be of sufficient magnitude to offset the combined force of these seven Biblical arguments for a geographically universal Flood in the days of Noah."[55] Central to their rejection of mainstream geology was their repudiation of uniformitarianism—the idea, stemming from James Hutton (1726–1797) and Charles Lyell (1797–1875) and embraced by Darwin—that one could uncover the mysteries of nature's past by observing its present processes. Denying uniformitarianism meant a simultaneous rejection of the knowability of geological history through naturalistic means. Because science was based on observation and testing, and geology—especially in its uniformitarian methodology—was purely historical, "it thus is impossible ever to prove that [past events] were brought about by the same processes of nature that we can measure at present." After all, "the only human observers—Noah and his family—recorded that the events were catastrophic!"[56] Whitcomb and Morris affirmed the biblically based "catastrophism" of George McCready Price as an alternative. Creation, in particular, and geology, in general, were therefore a black box—inaccessible to the tools and strategies of modern science and understandable only through the lens of scripture.

When compared to early anti-evolution of the Bryan era or to intelligent design in the late twentieth century, the flood geology of Whitcomb and Morris seems to be a world apart. Other than a general antipathy to Darwin and the supposed materialism of his worldview, there was far less ideological overlap than might at first appear. To the extent that these worldviews were "creationist," it could only be broadly so, in the sense that Bryan, Morris, and later ID advocates believed that God created the world. But such a definition of creationism is so generic as to be fairly unhelpful to the historian. Take, for example, the declaration by Theodosius Dobzhansky (one of the modern synthesis's aforementioned founders) of his own religion: "I am a creationist *and* an evolutionist. Evolution is God's, or Nature's, method of Creation. Creation is not an event that happened in 4004 BC; it is a process that began some 10 billion years ago and is still under way."[57] If creationism is defined only as the belief that God created the world, then it would include every worldview in existence other than only the most embittered atheistic materialism, and so its utility in many of these situations is debatable. In this sense, creationism equates only to theism—or perhaps even deism. This kind of creationism would not pass muster for the young-Earth creationists and flood geologists. For them, creationism was a total worldview—all-encompassing in its rejection of evolution and its affirmation of the Bible. To defend it, science must be reconstructed from the ground up and liberated from its secular trappings by basing creation in holy writ.

Genesis without God: The Alternative of Scientific Creationism

In the post-*Scopes* years, evolution had not factored significantly into high school education, but when it resurfaced as part of a national reawakening in scientific literacy, creationists of all stripes found that they could not keep it from being taught any longer. The Darwinian interregnum ended in 1968, when, in the *Epperson v. Arkansas* court case, a 1928 Arkansas statute (modeled on Tennessee's Butler Act, the source of the *Scopes* controversy) that prevented the teaching of evolution in public schools was ruled unconstitutional. The last vestige of the

Scopes fiasco was overturned; creationists were now forced to rethink their strategy. When blocking evolution from curricula was no longer possible, they resolved to earn creationism a space next to it. The way to achieve this was to make creationism suitable for secular instruction, to make it scientific but not religious. Explicitly biblical flood geology was modified; references to the Bible were either elided or softened, and an emphasis was put instead on what was called the "model" of "scientific creationism." This strategy was driven by legal reasoning as much as it was religious or philosophical, if not more so.[58] Henry Morris admitted this strategy openly, writing in the preface to the textbook *Scientific Creationism* that "most creationist books treat the subject of origins from the Biblical point of view, as well as the scientific, and, therefore, are not appropriate for instructional purposes in the public schools." The textbook, then, would fill a much-needed niche in that it sought to treat "all of the more pertinent aspects of the subject of origins and to do this solely on a scientific basis, with no references to the Bible or to religious doctrine."[59] Thus, creationism could be taught alongside evolution via a "two-model" approach.[60]

The "two-model" strategy alarmed and incensed critics, who interpreted scientific creationism as the most duplicitous form of anti-evolution yet faced. The paleontologist Niles Eldredge, one of the primary theorizers behind punctuated equilibrium, denounced scientific creationism as a "wolf in sheep's clothing"—a dangerous and destructive idea, inimical to all scientific inquiry, but one that was nevertheless attempting to pass itself off as legitimate science.[61] The philosopher of science Philip Kitcher dismissed it as "Genesis without God."[62] And according to the evolutionary biologist Jerry Coyne, "Creationists tried a new strategy: cloaking themselves in the mantle of science. This produced the oxymoronic 'scientific creationism.'"[63]

Nor were scientists alone in finding this strategy unpalatable. John C. Whitcomb himself was skeptical of the Bible's demotion in the world of scientific creationism, and he criticized the movement as "devoid of theological identity from a Christian perspective." In an attempt to garner respectability from secular scientists, Henry Morris and Gary Parker argued that the topics of creation, evolution, catastrophism, and uniformitarianism could "be evaluated strictly as scientific models, without reference to their theological, philosophical, or moral implications." Indeed, they added, "not all creationists believe in a personal God." For Whitcomb, this approach betrayed the religious purpose of creationism—without the Bible, what did it offer? "Can creationism retain its full power and beauty if it sheds its theological garments? By avoiding any mention of the Bible, or of Christ as the Creator, we may be able to gain equal time in some public school classrooms. But the cost would seem to be exceedingly high, for absolute certainty is lost and the spiritual impact that only the living and powerful Word of God can give (Heb 4:12) is blunted." Whitcomb worried that creationism, unmoored from the Bible, would drift far from safe harbor and lose truth in favor of expediency. "The brilliantly illuminating creation message," he wrote, "is a vital part of Biblical revelation—but it is an incomplete part in and of itself. Men desperately need the good news, not just more light."[64]

Creationism shorn of its biblical trappings—this accusation would again be leveled at intelligent design a decade after Whitcomb and Kitcher first decried scientific creationism with it. In ID, Whitcomb's fears would be realized, for the new movement would be even less committed to biblical foundations than scientific creationism was. With the Bible thus soft-pedaled, Noah's flood and the young Earth were likewise discarded like seeds by the wayside; even more worryingly, anti-evolutionists began to adopt heretical ideas like the validity of uniformitarianism and even common descent, surely fatal to any young-Earth form of creationism. However, while ID adopted these significant changes, scientific creationism still informed how it adapted. During the 1970s and '80s, new advances in biochemistry and molecular biology led creationists to grapple with evolution on a different front, and those who did so ended up laying the groundwork for ID's eventual emergence, and departure, from scientific creationism.

The Biochemical Turn

The Eagle pub in Cambridge rests across the street from King's College, just to the south of Great St. Mary's via the King's Parade road. Today, it is so swarming with visitors that all its quaint Oxbridge charm is smothered by lanyard-bearing tour groups; in 1953, the pub was quieter and, due to its proximity to the Cavendish Laboratory, was a frequent meeting place for scientists at lunch. On the 28th of February, Francis

Crick (1916–2004) burst into the pub and proclaimed out loud to all the patrons that he and his American colleague James Watson had discovered "the secret of life."[65] The discovery of the double helix, the shape of DNA, reverberated throughout the world, becoming a landmark in the history of molecular biology and in human understanding of the language of life. The discovery of DNA was only one of two major events that year, both of which would dominate the rapidly growing biochemical field. Across the pond, an experiment was undertaken that would frame the conversation about the origin of life for the rest of the century. Only three months later, in May 1953, a paper on chemical evolution (the primordial evolution of chemicals into complex molecular systems) was published in *Science*. The chemistry graduate student Stanley Miller (1930–2007), working in the University of Chicago lab of Harold Urey (1893–1981), synthesized amino acids from inorganic compounds by circulating several molecules past electric discharge, following ideas developed by Alexander Oparin (1894–1980) in 1924. In simulating the early Earth's atmosphere (thought to be composed of "methane, ammonia, water, and hydrogen instead of carbon dioxide, nitrogen, and oxygen, and water"), Miller had shown how the basic building blocks of proteins, amino acids, could have spontaneously arisen from inorganic matter.[66] With both the shape of DNA and the emergence of proteins demystified, the resolution of abiogenesis—of the origin of life—was tantalizingly within grasp.

Darwin's *Origin of Species* famously avoided the origin of life. As with the mechanism of heredity, the possible emergence of life from inorganic matter was purely speculative. "But if (and oh what a big if) we could conceive in some warm little pond," Darwin wrote in a letter to J. D. Hooker, "with all sorts of ammonia and phosphoric salts—light, heat, electricity etcetera present, that a protein compound was chemically formed, ready to undergo still more complex changes."[67] Darwin's speculation found empirical confirmation in the experimentation of his twentieth-century followers—from Oparin and Haldane to Miller and Urey. The critical mass of these achievements took place in the midcentury, nearly one hundred years after the *Origin*. "As we know it today," writes Edna Suarez-Diaz, "the study of evolution at the molecular level owes its modern face to the rise of protein chemistry, biomedicine, and molecular biology after the Second World War."[68] During this time of revolutionary discovery, when a plausible mechanism for a nonmiraculous origin of life was given laboratory support, creationists—who had before the 1950s paid relatively little attention to biochemistry and the nascent science of molecular biology—found that the war against evolution had expanded to a new front.

Despite the stunning achievement of the Miller experiment, a great many unsolved and unanswered questions remained concerning the origin of life. The situation even today has not improved as much as scientists had hoped. "The hiatus between the primitive soup and the RNA world [a popular hypothesis about the origin of life] is discouragingly enormous," writes the biologist Antonio Lazcano. Because "we will never know how life first appeared," any investigation must be "inquiring and explanatory rather than definitive and conclusive."[69] That does not mean that there has been little effort to discover how life might have originated. In the midcentury, and following the early developments in the field, the biophysicist Dean H. Kenyon, professor of biology at San Francisco State University, coauthored with Gary Steinman a book titled *Biochemical Predestination*. Released in 1969, the book sought to be a compendium of research success and failure in the field of chemical evolution, specifically abiogenesis (the emergence of life from nonlife). "Our aim in the present book," they wrote, "is not only to present in detail what we consider to be the major experimental approaches to the origin of life problem . . . but at the same time to evaluate critically the *underlying assumptions*."[70] In the book's foreword, the famed biochemist Melvin Calvin (1911–1997) called *Biochemical Predestination* "really the first attempt to produce what might be called a comprehensive essay which could be used as the basic textbook for systematic discussion of the problem in an academic and scientific environment."[71] Chief among the assumptions for research into abiogenesis was the young-Earth creationist bugbear of uniformitarianism. Wrote Kenyon and Steinman, "It is universally assumed by everyone working in the field that the details of physical and chemical laws have remained essentially the same from the epoch of origin down to the present day." The reliable constancy of the Earth's processes—same today as yesterday and forever—must be accepted as a first principle "if the formulation of the origin of life problem as an essentially historical problem is to have any meaning." Furthermore, and this would have induced groans from creationist critics, "Implicit in this assumption is the requirement that no supernatural agency 'entered nature' at the time of the origin, was crucial to it, and then withdrew from history."[72] Instead of supernatural intervention, Kenyon and Steinman cautiously proposed the theory of "biochemical predestination," meaning that "there appears to be an inherent tendency toward the type of organization which we observe in living cells."[73] For Kenyon and Steinman, rather than the random-chance approach to the origin of life (exemplified by the Urey-Miller experiment, which operated under the assumption that fortuitous events in the distant past might by coincidence lead to life), they suggested that the inherent properties of nature might inexorably produce living matter. Life, in this instance, did not generate spontaneously but instead arose via intrinsic development. The ID supporter and philosopher of science Stephen Meyer wrote many years later that this proposal was an early version of what came to be called "self-organization."[74]

On the other side of the Atlantic, an organic chemist and young-Earth creationist named A. E. Wilder-Smith (1915–1995) came across *Biochemical Predestination* and became determined to refute it. Wilder-Smith had been raised a Christian but abandoned the faith and became an atheist when he decided that Christianity was "for the most part humbug and . . . hypocrisy." As he grew older, he was reacquainted with religion under the tutelage, first, of a general in the British army and then, later, C. S. Lewis (1898–1963). Wilder-Smith met Lewis while studying at Oxford and, according to Wilder-Smith's wife, Beate, "Lewis's superb lectures and books deeply impressed and greatly influenced [Wilder-Smith's] intellectual life and thinking. For many years, he committed himself to finding scientific answers to the genesis of life, the origin of genetic information and man."[75] In 1970, Wilder-Smith produced a creationist work titled *The Creation of Life: A Cybernetic Approach to Evolution*, noteworthy for its extensive discussion of the problem of abiogenesis and its full-bore attack on Dean Kenyon. He criticized Kenyon's uniformitarian methodology and "exclusion of the theistic *Weltanschauung*," as well as his conclusions, writing that "Kenyon himself does not seem, in his heart of hearts, to be very convinced by his own efforts."[76] He accused Kenyon of lacking metaphysical creativity and therefore limiting his own understanding of the origin of life's information (the DNA code). "Where," asked Wilder-Smith, "did the superorder of the algorithm come from?"[77] Rather than excluding the activity of a divine creator, as Kenyon had done, Wilder-Smith proposed an alternative: he turned to the old argument from design and to William Paley (1743–1805), the English cleric who had articulated the "watchmaker argument" so lucidly and popularly in his 1802 work *Natural Theology*. For Wilder-Smith, Darwin's discovery of natural selection—which had reduced the design of complex things in nature to only *apparent* design—did not apply at the biochemical level. With the discovery of the genetic code and the complexity of microbial life, "Darwin's erstwhile philosophical victim can live once more. . . . Paley is reestablished and once more the design [of cells] is found to throw light on the designer."[78] After a lengthy absence, the language of design—which Whitcomb and Morris did not rely on very heavily—had reentered the creationist lexicon and occupied center stage, not the periphery.[79]

Wilder-Smith could not have known at the time how correct he was about the doubts within Kenyon's own mind. Like Phillip Johnson after him, Kenyon's transition took place in England, a geographical change of scenery occasioning an ideological shift. "In 1974, I went to Oxford University as part of a sabbatical leave and spent the time reading and interviewing people on the relation between science and Christian faith," Kenyon later explained in an interview. "Then in 1976, a student gave me a book by A. E. Wilder-Smith, *The Creation of Life*. . . . Many pages of that book deal with arguments against *Biochemical Predestination*, and I found myself hard-pressed to come up with a counter-rebuttal." Kenyon entered a period of crisis. He read more creationist works, including *The Genesis Flood*, and became convinced.[80] Scientific creationists did not waste any time claiming the allegiance of their new convert, and only a few years later, Kenyon began publishing on creationism himself. In 1982, on the eve of scientific creationism's day in Arkansas court (in *McLean*), Kenyon wrote the foreword to Henry Morris and Gary Parker's *What Is Creation Science?* "I no longer believe," wrote Kenyon, "that the arguments in *Biochemical Predestination* . . . add up to an adequate defense of the view that life arose spontaneously on this planet from nonliving matter." Kenyon disputed the notion that scientific creationism was just religion in disguise and called for open-minded inquiry. For any interested, *What Is Creation Science?* was the best book available, and "if after reading this book . . . [one] still contends that the creationist view is religion and the evolutionary view is pure science, he should ask himself whether something other than the facts of nature is influencing his thinking about origins."[81] Here we see the later ID language of worldview schism present, but that would not be the only anticipation of ID thought in the book.

Gary Parker, who composed the first half of *What Is Creation Science?* (the second half was Morris's), laid out various arguments that would later become strongly associated with intelligent design. He contended that the origin of life—and especially DNA and proteins—was an insuperable problem for chemical evolution, just as Wilder-Smith did and Stephen Meyer later would.[82] He argued that the human eye was too complex to have arisen gradually and cited the same damning excerpts from Charles Darwin, Richard Lewontin (1929–2021), and Stephen Jay Gould (1941–2002) that both Phillip Johnson and Michael Behe later would.[83] He also attacked the fossil record, which he saw as discontinuous, and cited the plaintive cries of "saltationists" like Richard Goldschmidt (1878–1958), who longed for a "hopeful monster" to explain macromutation, as proof that evolution was in trouble as a theory, just as Johnson would in *Darwin on Trial.*[84] In Parker's writing, one finds an adumbration of many of the later concepts associated with intelligent design—the lines of argument gaining clarity. At one point, Parker chronicled the vexation that struck the biologist Garrett Hardin (1915–2003) when the latter studied the seemingly intentional adaptations of living creatures to their environment. Hardin opined, "Is the [evolutionary] framework wrong? . . . Was Paley right?" Here, the bridge from young-Earth creationism to ID was being built. Focusing on design like this was a rather unusual angle for youth-Earth creationists, and Parker, true to his self-effacing style, wrote, "If you're like me, you never heard of William Paley. . . . Paley was a thinker in the 18th century who argued that the kind of design we see in the living world points clearly to a designer." He asked again, "'Was Paley right?' That is, can we infer creation from the kind of design we see among living things?"[85]

Parker left the question rhetorical, but what can be seen here was a gradual opening within creationism to the arguments that would become the quintessence of ID. However, while Parker started on a path toward more ID-esque attacks on evolution, he did not abandon the young Earth or the need for a global flood, as later IDers would. What his work shows is that once young-Earth creationists began detaching their biblical arguments from their scientific ones (as the legal and educational strategy of scientific creationism necessitated), elements of proto-ID theory can be found. Gary Parker came close to the edge of design—to a bygone world of natural theology versus Darwin without dependence on worldwide floods or Adam and Eve's predilection for fruit. The leap remained unmade, though, and the second half of *What Is Creation Science?*, Morris's half, contained the familiar creationist arguments for the global flood, for the decay of the speed of light, and for the incompatibility of evolution with the second law of thermodynamics—all creationist mainstays that would have a much harder time maintaining favor in the world of ID. However, Parker and Morris's work indicated an increasingly bifurcated tension in the world of scientific creationism and an emerging gap between the arguments derived solely from science and those whose foothold was still in the world of the Bible and the Genesis narrative. In the ensuing decade—the 1980s—the split would become even more pronounced, and in the aftermath of two disastrous court cases, ID would more fully separate from scientific creationist ideology.

Baptism by Fire: The Legal Defeat of Scientific Creationism and the Birth of Intelligent Design

"If this stuff is science," Niles Eldredge recalled the presiding judge in *McLean v. Arkansas*, William Overton (1939–1987), asking, "why do we need a law to teach it?"[86] The question of whether creationism was science or not dominated 1982's *McLean* and ultimately resulted in defeat for the creationist side. The case was occasioned by the new creationist strategy of bringing creationism into public schools under the guise of science, no religion added. An Arkansas state law, Act 590, was passed in 1981, mandating the teaching of creationism along with evolution. Educators disputed the act's constitutionality, and the court case resulted. It must be noted that the titular McLean of the plaintiffs was not a frothing-at-the-mouth atheist political crusader but rather the Reverend William McLean, a United Methodist minister. The argument was largely set by the philosopher of science Michael Ruse, who not only testified to scientific creationism's scientific inaccuracy but also argued that it was not science at all. Overton followed Ruse's explanations closely in his ruling, determining that scientific creationism "is simply not science." Furthermore, creation science had as its core impetus a desire for the "advancement of religion," which could not plausibly be excused as secondary if it had no real scientific merit. This was a violation of the so-called *Lemon* test, stemming from the 1971 *Lemon v. Kurtzman* US Supreme Court decision. As Overton wrote, "The second part of the three-pronged test for establishment reaches only those statutes having as their *primary* effect the advancement of religion. Secondary effects which advance religion are not constitutionally fatal. Since creation science is not science, the conclusion is inescapable that the *only* real effect of Act 590 is the advancement of religion."[87] Thus, guilty of violating that most hallowed of amendments, scientific creationism was struck down on establishment clause grounds (but only in Arkansas, for *McLean* never went beyond the district level).[88]

The defense for Arkansas was flummoxed from the start. Of chief importance for the creationist side was the lawyer Wendell R. Bird, who continued a vaunted trend of lawyerly opposition to Darwin: from George Curtis (1812–1894) to William Jennings Bryan to Phillip Johnson. Bird, a graduate of Yale Law School, had been a theistic evolutionist when younger but become a creationist later in life. He formulated a new strategy to make war against evolution, this time by, in Edward Larson's words, "reviv[ing] the well-worn argument that teaching only evolution violated the free exercise of such religions" that, as Bird said, "affirm divine creation as a cardinal tenet of faith."[89] In the case, Bird became frustrated by the Arkansas defense of Act 590, which he interpreted as noncommittal and disinterested. The defense blocked Bird and assured him that, as Larson chronicles, the state attorney general would "defend the suit with adequate vigor and diligence."[90] Unconvinced, Bird continued to meddle in the case. The day before Dean Kenyon's scheduled testimony (in which he would explain that it was scientifically licit to hold that life's origin on Earth required supernatural assistance), Kenyon vanished without a trace. He had apparently flown the coop, left Little Rock, and gone back to California; it turns out, Kenyon did so at the urging of Bird, who apparently wanted to "save" him for a better legal opportunity in Louisiana, where there was a similar balanced-treatment act under scrutiny.[91] Steve Clark, the Arkansas attorney general, considered legal action against Bird for behavior he considered "tantamount to tampering with justice," but he did not pursue it.[92] Even had Kenyon testified, it may not have amounted to much—the defense struggled to find expert witnesses able to show that creationism was truly science. One witness, the Christian philosopher Norman Geisler, embarrassed the defense because he had earlier "declar[ed] that UFO's were agents of Satan," and, as observed by *Science* magazine, the computer scientist Henry Voss "was rapidly withdrawn" by the state when "he too began to expound on things satanic and demonical."[93] Creationists were frustrated by this. Later, Wayne Frair maintained that the ACLU's lawyers brought up Geisler's pretrial comments about alien spaceships to discredit his testimony on science.[94] However, even without Kenyon, the defense still sought to demonstrate the implausibility of abiogenesis and the viability of creationism as an alternative, but the terminal embarrassment came when the state's "star witness," Chandra Wickramasinghe, testified on this matter. Wickramasinghe, an astronomer and frequent collaborator with the legendary English astrophysicist Fred Hoyle (1915–2001), was summoned by the creationist side to testify regarding the difficulties of abiogenesis, which he did. But he then went further by rejecting the basic concept of creationism, saying that "no rational scientist" could believe in a young Earth.[95] After creationism's defeat at *McLean*, there was dissent within the creationist ranks on how to proceed. Geisler had wished that *McLean* would be appealed, but his co-creationists, such as Jon Buell at the Foundation for Thought and Ethics and Wendell Bird himself, switched focus to Louisiana and the case that would eventually make it to the Supreme Court in 1987: *Edwards v. Aguillard*.[96] This case, however, would provide the same result for creationists, but this time with nationwide consequences. Deploying the *Lemon* test just as Overton did, the majority opinion, under guidance by Justice William Brennan (1906–1997), concluded that the purpose of Louisiana's Balanced Treatment Act was, primarily, to disseminate religion and thus violated the First Amendment.[97]

In the aftermath of *McLean* (and, later, *Edwards*), anti-evolutionists struggled to regroup and rebrand. In the spirit of the Apostle Paul, the court loss provided an opportunity to "test all things and hold fast to what is good." What was good appeared to be an emphasis on problems in abiogenesis and biochemistry and an avoidance of biblical and doctrinal issues. In 1984, the Foundation for Thought and Ethics redoubled its efforts and produced a critique of chemical evolution with an old-Earth flavor, *The Mystery of Life's Origin*. Dean Kenyon composed the foreword and contended that neither experiments along the lines of Miller-Urey nor research such as his own *Biochemical Predestination* were viable for origin of life studies. In his view, none of the proposed solutions were promising. This conclusion would be unpopular, noted Kenyon, for scientists feared that "acceptance of this conclusion would open the door to the possibility (or the necessity) of a supernatural origin of life."[98] Ronald Numbers argued that it was Kenyon's foreword that was the most important aspect of the book, while Thomas Woodward contended that it was its epilogue, which suggested an intelligent creator as a potential cause of the origin of life on Earth (among four other alternatives).[99] In the authors' words, "special creation by a creator beyond the cosmos" provided a solution to the problem of abiogenesis. Some of the issues with this were then run through: the authors agreed that "miracles must not be posited for operation science" (which they distinguished from "origin science"), but they also disputed that "the exclusion of the divine from origin science is valid. This has not been demonstrated." The three authors, Walter L. Bradley, Roger L. Olsen, and Charles B. Thaxton, went on to argue that the banning of any possible divine activity from origin of life studies resulted not from scientific thinking but rather from a commitment to "metaphysical naturalism." In lieu of this philosophical perspective, the authors then issued a call for "metaphysical tolerance" regarding potential creation of the first life.[100]

Mystery departed from young-Earth creationism in a few key ways. It contained no overt or implicit biblical references or attempts to establish doctrines like the Fall of Man or original sin. It dealt not at all with Darwinian evolution. It had no interest in the flood. Critical for their methodology, the authors did not wish to simply discard uniformitarianism altogether as a scientific working assumption, as Morris, Whitcomb, and even Kenyon had done. They wrote, "The developers of chemical evolution theory acknowledge its speculative nature. . . . We agree that there is scientific value in the pursuit of such reconstructions that should not be dismissed out of hand."[101] Without a young Earth, with a focus only on chemical evolution, with a toleration of uniformitarian principles, and with an absence of any explicit religious doctrine, *Mystery* represented a shift in anti-evolutionist thinking, strategy, and ideology. It would have a great effect on later intelligent design advocates, particularly Stephen Meyer, who credited Thaxton personally.[102] Meyer contended that *Mystery* was the true origin of intelligent design thinking, writing that it "was not developed as a legal strategy, still less as one to abet creationism."[103] Because *Mystery* antedated the 1987 *Edwards* case, Meyer argued that the book's ideas could not be dismissed as simply a legal sidestepping of the judicial issues caused by creationism's explicit religiosity (though he did not remark on the creationist Kenyon's presence or that 1984 was two years after the *McLean* case). Woodward, too, claimed that *Mystery* did not fit the "genre" of scientific creationism.[104] However, even if *Mystery* was not explicitly creationist in the same manner as *What Is Creation Science?*, its intellectual lineage and debt were explicit, and not only because Kenyon wrote both forewords. The origin of life critique—the focus on the creation of information, of DNA, in the cell—had been rapidly growing in the creationist world, finding advocates in Wilder-Smith, as well as the creationist super-soldier Duane Gish (1921–2013) and more fringe writers such as Randy Wysong.[105] What *Mystery* did was invigorate creationism by targeting a particular weak point in the cosmic evolutionary picture and insulating itself against criticism by removing religious concepts relating to the Bible and Christian dogma, dispensing with the focus on the young Earth and worldwide flood, and embracing uniformitarianism.

Geisler too leaped back into the fray, this time in his follow-up book to *McLean*, titled *Origin Science*. Strongly influenced by *The Mystery of Life's Origin* (the foreword to Geisler's new book was written by Walter Bradley, and Geisler thanked Thaxton, Bradley, and Kenyon in his acknowledgments), Geisler argued for something of a reboot after the *McLean* and *Edwards* cases. What is most interesting about this book is its advocacy of terminology and concepts that would later become central to ID. Geisler repeatedly turned to "specified complexity," arguing that the kind of complex information in DNA could only have been created by an intelligence—similar to the arguments that William Dembski and Stephen Meyer would later make.[106] He anticipated Michael Behe when he contended that new developments in molecular biology resurrected William Paley's design argument.[107] He embraced the Big Bang theory (a literally cosmic departure from young-Earth creationism) and even used the phrase "intelligent designer" in discussing Charles

Lyell.[108] And he cited Michael Denton, one of ID's progenitors.[109] However, there are important differences between Geisler and ID. Though he was more open to uniformitarianism, Geisler stopped short of applying it to questions of "origin." For Geisler, science must be split into two forms: origin science and operation science. Operation was the everyday kind of science we all know, but origin investigated "singularities"—one-off events that had no parallel and that therefore could not be studied under regular scientific practice.[110] Treating creation events—like the origin of the first cell—as black boxes would prove to be a vestige of scientific creationism that ID advocates such as Stephen Meyer would not maintain. ID would fully embrace uniformitarianism and then try to use it to undercut Darwin. That said, Geisler's book is important in highlighting the transition between scientific creationism and ID.

Wendell Bird undertook a similar strategy of critique and rebrand. In a massive, two-part tome titled *The Origin of Species Revisited*, Bird argued against evolution on all fronts. Rather than advocate for creation or creationism, however, Bird chose to use the phrase "abrupt appearance" to represent his general stance of anti-evolution that spanned across various scientific disciplines. The change was more than superficial, for Bird began adopting more consciously old-Earth arguments and anticipated ID ones. Throughout the two books, Bird covered "abrupt appearance" regarding abiogenesis (framing the "sudden" creation of the cell and DNA), the origin of the universe (endorsing much of the science of the Big Bang and the "anthropic principle"), and the origin of megafauna (discussing the "abrupt appearance" of major species during the Cambrian explosion).[111] Advocates of abrupt appearance meant for it to span the spectrum from biochemistry to astrophysics to paleontology; gone were the pitfalls of young-Earth creationism and the millstone of Genesis around its neck. Bird relied heavily on Denton's 1985 anti-Darwinist work *Evolution: A Theory in Crisis*, which not only focused on the problem of life's origin but also revived the old, nineteenth-century typological argument against the endless malleability of species.[112] Denton, of course, was one of the primary forerunners of intelligent design. Describing himself as an agnostic, Denton attacked Darwinism with no explicit religious angle (though critics alleged that there was an implicit one).

Along with Denton, the lawyer Phillip Johnson read Bird as well and reviewed *The Origin of Species Revisited*. Johnson noted that, at *Edwards*,

Bird "labored unsuccessfully to rescue scientific creationism from its fatal association with Biblical fundamentalism."[113] To achieve this herculean task, an anti-evolutionist must decouple creation from Genesis, science from biblical literalism, and coin a new name. Johnson wondered whether a "purely descriptive" name like "abrupt appearance" fit the bill, but he appreciated its effort to disentangle the knotted skein of creationist, anti-evolutionist, and fundamentalist ideas. Wrote Johnson,

> In short the descriptive phrase "abrupt appearance" accurately characterizes the first appearance of animal groups, if one looks at the evidence without theoretical preconceptions. . . . Of course the abrupt appearance in question is thought to have occurred hundreds of millions of years ago, and not in the year 4004 B.C. . . . The fundamental claim of religious creationism, however, is simply that God creates, and this claim has no necessary connection to any particular theory about Biblical authority, however much it may be associated in practice with Biblical literalism. Creationism in the broadest sense is not even hostile to "evolution," provided that evolution is considered to be a process through which God creates.[114]

Johnson saw in Bird's treatise a possible path forward—a deemphasis on the biblical nature of the argument, a retreat from the word "creationism," and a focus on attacking the evolutionary world picture wherever any weaknesses could be found while softening the need for an alternative. Perhaps "abrupt appearance" would not suffice—so what would?

Dean Kenyon was not yet finished. While the *Edwards* case was ongoing, and while Johnson was reading Denton and undergoing his own conversion to anti-evolution, the Foundation for Thought and Ethics hired Kenyon to write a new textbook that would critique evolution without explicit religious ideology. Such a book would therefore be suitable for public schools, just as *Scientific Creationism* was meant to be. Writing with Percival Davis, Kenyon focused on critiquing evolution from the vantage of cellular complexity, chemical evolution and abiogenesis, and the changeability of species. During its drafting, the book cycled through various names, such as *Biology and Creation* (1986) and *Biology and Origins* (1987), before finally landing on the innocuously Robert Burnsian title *Of Pandas and People*. In the aftermath of *Ed-*

wards, the words "creation" and "creationist" were taboo, threatening to doom any anti-evolution effort from the start. Instead, at the prompting of Thaxton, Kenyon and Davis alighted on a substitute name, one that they hoped would not bring with it the baggage that had burdened creationism throughout its legal and political failures in the 1980s. They chose "intelligent design."[115]

Conclusion

By the year 1990, the good ship creation had been finally scrubbed clean of all the barnacled accretions that had glommed onto its hull over the years: a young Earth, a global flood, a garden inhabited by "the two orchard thieves," as Herman Melville called them. All these features were expunged from the design-focused worldview, and the emergence of intelligent design represented, at least for a time, the victory of science over the Bible in the world of anti-evolution. Further, that these anti-evolutionary arguments became less burdened by biblical baggage meant that even though they had their sources in creationist literature, they could appeal to skeptics of Darwin who did not have creationist backgrounds: figures like Phillip Johnson, Michael Behe, and William Dembski. The dislodging of explicitly biblical concepts from anti-evolution was only the first step in a longer process of schism between reason and revelation. As for creationism, ID's relationship to it was complicated and at times murky, but the characterization of both its supporters and its critics is only part of the story. Critics wished to unveil ID's creationist heritage; advocates attempted to separate ID from creationism by implying that its origin was separate and that it was concerned with different questions. Neither picture is completely accurate. ID, though not identical to creationism, was not removed either. It draws on multiple creationist strains and is a hybrid of old nineteenth- and early twentieth-century critiques along with biochemical and molecular arguments from the mid- to late twentieth century. But ID became its own movement as the anti-evolutionists who supported it gradually shed, one by one, the biblical and religious accoutrements that had dragged scientific creationism into the mud and smothered it under their own weight. This new form of anti-evolution would, its proponents hoped, have no such problems. In fact, many creationists greeted ID with unease due to its theological differences from other forms of anti-evolution. The question remains, though: What was intelligent design? What was new about it? What were its metaphysical, scientific, and theological beliefs?

In investigating these elements of ID, we will see that its intellectual content was more rigorous than most creationist writings, and its proponents were more credentialed and experienced in academic and professional science. ID was, at bottom, an intellectual movement based on ideas—and ideas, as Richard Weaver was convinced, have consequences. What those consequences will be, however, is not always obvious. Underneath ID's focus on design and Darwin were radical notions that challenged the very practice of science and the reliability of human knowledge acquisition. ID staked out bold positions on how science should be done, on what impact metaphysics should have on science, and on what the role of theology should be in these fields. Such perspectives would, over time, engender deep suspicion and hostility toward the scientific establishment and the government's policies on science and education. It did not start out this way, but buried within its ideological structure were ideas that would unfold into a radical revision and then rejection of mainstream science.

2

Design

The Beginning of the Intelligent Design Movement

Phillip Johnson became interested in refuting evolution while on sabbatical in England in the late 1980s. Up to that point, Johnson had been an accomplished law professor at the University of California, Berkeley, author of several books on criminal law, and a former law clerk for Chief Justice Earl Warren (1891–1974). In the middle of his life, he went through a divorce and then a conversion to evangelical Christianity, which triggered a desire to change his priorities. While in England, he read the zoologist Richard Dawkins's argument against design, *The Blind Watchmaker*—Johnson felt it was "fairly convincing on the first reading but full of holes on the second"—and then Michael Denton's *Evolution: A Theory in Crisis*, which "did much to alert [him] to the issues." He began to question the received wisdom of evolutionary biology, but the great lightbulb moment for Johnson was when he and Kathie, his second wife, were browsing a London bookstore. Kathie pulled up the legendary sci-fi writer Isaac Asimov's (1920–1992) *Guide to Science*, whose entry for evolution was only a "brief" description followed by "three pages of heavy-handed ad hominem denunciation of creationists for not accepting the absolute truth of this theory that was so obvious to all thinking persons." When they looked at the "proof of evolution" section, it contained only the peppered moth experiment (a 1950s study on the appearance of dark-colored moths in industrial Birmingham—demonstrating Darwinian fitness of the dark-colored trait as it provided camouflage in the newly polluted woods). Upon reading this, Kathie remarked, "I think you're on to something."[1]

Johnson decided to devote his life to refuting Darwinism. First, he enmeshed himself in studying evolutionary biology and composed a paper in 1988 titled "Science and Scientific Naturalism in the Evolution Controversy." After sending the paper to the staunch atheist and historian of science William Provine (1942–2015), Johnson discussed it in a 1988 Faculty Seminar at Berkeley, with the evolutionary biologist Montgomery Slatkin attending (at the suggestion of Provine).[2] In the paper, Johnson immediately dismissed young-Earth creationism and rebranded the struggle—it was not evolution versus creation but rather naturalist versus theistic metaphysics. Johnson described creationism as biblical literalism and the belief that God created "basic kinds" of species "within the space of a single week about six thousand years ago," a position he emphatically rejected. As he later summarized in a follow-up review, "I have no interest in promoting or even discussing creationism in that sense." Instead, he said, "The question I raise is not whether science should be forced to share the stage with some biblically based rival known as creationism, but whether we ought to be distinguishing between the doctrines of scientific materialist philosophy and the conclusions that can legitimately be drawn from the empirical research methods employed in the natural sciences."[3] Johnson argued that an assumed atheistic worldview animated evolutionary biology and compelled its adherents to accept evolution because it was the only plausible naturalistic explanation. He documented admissions of naturalistic sympathies from the evolutionary biologists George Gaylord Simpson (1902–1984), Douglas Futuyma, Richard Dawkins, and William Provine to illustrate that naturalism was the key to evolution's success. Dawkins's claim that "Darwin made it possible to be an intellectually fulfilled atheist" neatly encapsulated Johnson's key thesis.[4] Thomas Woodward wrote that Montgomery Slatkin sent Johnson a letter urging the law professor to publish his critiques of evolution in a sequence of academic journals, lest he be appropriated by creationists seeking a new leader in anti-Darwinism. When Woodward inquired about this, Johnson relayed that he was "amazed" by the suggestion that, as Woodword summarized, he "ought to quarantine himself from the rabble of popular creationism and immediately started rewriting his paper into book form."[5] (One sees here a bit of the anti-establishment mindset that would characterize much of the ID movement). That book would eventually be released in 1991 as *Darwin on Trial.*

For ID's supporters, such as William Dembski, it was "hard to overestimate the impact" of *Darwin on Trial.* After years of the disjointed, sometimes amateurish, and scientifically unsuccessful critiques proffered by creationists, Johnson created the "beachhead that finally put effective criticism of Darwinism on the map."[6] Johnson's protégé, the creationist philosopher John Mark Reynolds, wrote that *Darwin on Trial* was a "seminal book" but one that admittedly did not bring much that was new to the debate. Critics argued that it only repeated earlier claims made by creationists, though it was better packaged and shorn of any overt biblical connections. Reynolds contended that this was the great virtue of *Darwin on Trial*, for before Johnson, "most of the arguments had not been drawn together into one place in an intellectually respectable manner." In a refreshing strategy for anti-Darwinists, Johnson ignored "side issues" and disregarded "sectarian [debates that did not] appeal to a wide audience" (at first, at least—he later became more overtly religious in his published writing).[7] Though the book's content was important, Johnson himself loomed large in the story of ID as a unifying personality. The geologist and ID legal advocate Casey Luskin called him the "Godfather" of intelligent design.[8] Reynolds compared the anti-evolutionary cause to the Council of Elrond from *The Lord of the Rings* and noted that Johnson would have been invited to a real-life version (Reynolds did not forget Tolkien's fellow Inkling C. S. Lewis and also described Johnson as Elwin Ransom, the hero of Lewis's Space Trilogy).[9] Johnson served as the figurehead when it came to the ID strategy to unseat Darwin.

One important part of this story—the story of populism and the development of ID as a sort of intellectual counterculture—is the emergence of the internet and the way ID advocates used it to their advantage in the early 1990s. This was still a new and unusual technology for most Americans at the time, but in academic circles, it was becoming widespread. As Dembski recollected, one of Johnson's most significant acts was to "insist that the participants [of Pajaro Dunes, an early ID conference] get on e-mail and be part of a listserv that he would run from Berkeley." The significance of this move can be hard to understand now. "Today, everyone has e-mail," wrote Dembski. "That was not the case in 1993." Netscape and Internet Explorer were not commonly used yet. "But Phil saw what was coming," he continued, "and how the Internet would allow for the dissemination of knowledge that would make it increasingly difficult for secular elites to maintain control over what people think."[10] Michael Behe was an early beneficiary of ID's organization on the internet. As he wrote, "With the aid of the then newfangled internet, over the years I met other academics who had had experiences roughly similar to mine, who had been perfectly willing to accept Darwinian evolution, but at some point realized with shock that the larger theory was an intellectual façade."[11]

The democratizing impulse of the internet was built into it from the beginning, and some of its early visionaries—especially J. C. R. Licklider ("computing's Johnny Appleseed")—had great hope in the network of networks to assist in human thinking. Licklider's prophecy for computers was that "they would democratize access to information, foster wider communities, and build a new global commons for communication, commerce, and collaboration."[12] Licklider's paper on the "Intergalactic Network" would, as the physicist M. Mitchell Waldrop explains in his history of computing, "become the direct inspiration for the Arpanet, which would eventually evolve into today's Internet."[13] IDers did indeed use the internet in such a manner, but they did so to form a sort of scientific insurgency. It had a democratizing impulse, to be sure, but perhaps not the kind of knowledge building that the internet's past luminaries would have anticipated. The internet's capacity to support anti-establishment and anti-authority enterprises is critical to the story of ID.

To understand the significance of both Johnson and *Darwin on Trial*, one must look not to his science but to his philosophical critique of naturalism. For Johnson, naturalism was the basis for Darwin's victory, and so to demonstrate his inadequacy, one must reveal evolution's philosophical foundation. After an examination of the philosophy of design, the role that science played in ID would be easier to understand, for ID's adherents hoped to use science to break down the philosophical edifice of the Darwinian paradigm. The third leg of design, theology, was more controversial for anti-evolutionists, with most ID advocates distinguishing themselves from previous opposition by, in theory, resisting the links between theology and science (even natural theology), doctrine, and scripture. In practice, this separation proved challenging.

The Philosophy of Intelligent Design: Metaphysics Matters

Phillip Johnson was not the first anti-evolutionist to locate the core of the battle in metaphysics, but he was arguably the most prominent academic, with the necessary scholarly credentials, to do so; and he also focused the argument by making it his lynchpin. On the first page of *Darwin on Trial*, Johnson tried to reset the game by defining creation as essentially any view that God exists and creates. It was incompatible only with naturalism. "'Evolution' contradicts 'creation,'" he wrote, "only when it is explicitly or tacitly defined as *fully naturalistic evolution*." This clarification would, Johnson hoped, clear up the confusion regarding whether all anti-evolution was young-Earth creationist.[14] "Darwinists," he contended, "identify science with a philosophical doctrine known as *naturalism*" (for Johnson, this was synonymous with materialism). He continued, "Naturalism assumes the entire realm of nature to be a closed system of material causes and effects, which cannot be influenced by anything from 'outside.'"[15] Naturalism necessitated Darwinian evolution, because when naturalism was assumed in science (as, for Johnson, it almost universally was), then the physical world's "features must be explicable in terms of forces and causes accessible to scientific investigation. It follows that the best naturalistic explanation available is effectively true, with the proviso that it may eventually be supplanted by a better or more inclusive theory."[16] Because Darwinian evolution was the best naturalistic explanation, it was embraced with a diabolical fervor. When Stephen Jay Gould blasted Johnson in a review of *Darwin on Trial*, Johnson responded, "What divides Gould and me has little to do with scientific evidence and everything to do with metaphysics. Gould approaches the question of evolution from a philosophical starting point in scientific naturalism."[17]

Johnson had keyed onto the idea of Darwinism-as-materialism when he read Richard Dawkins's *The Blind Watchmaker*. Dawkins went so far as to define biology as "the study of complicated things that give the appearance of having been designed for a purpose."[18] It might seem at first glance that life's complexity could not be explained without a designing intelligence—an intuitive and commonsensical idea—but for Dawkins, Darwin disposed of this issue by supplying the first plausible alternative for nature's "apparent design." Furthermore, Darwin revealed the world to be ontologically random and therefore pointless. For Dawkins, "the universe we observe has precisely the properties we should expect if there is, at bottom, no design, no purpose, no evil and no good, nothing but blind, pitiless indifference."[19]

Johnson accepted Dawkins's point about the fundamental incompatibility of Darwin and deity, as well as evolution and purpose.

Evolution and atheism were fundamentally linked, and Johnson characterized the modern evolutionary synthesis as the "blind watchmaker thesis," which meant that "there was no need for a Creator." Furthermore, this "blind watchmaker thesis" was critical to the entire apparatus of secular modernity and the liberal political order undergirded by it, for "this thesis is the most important element in the creation story of modernist culture—a story that is promulgated aggressively in the educational world and the media with the resources of government."[20] Dawkins, because of his position as a vocal proponent of atheistic naturalism and the gene-centered view of evolution (as espoused in his landmark *The Selfish Gene*), served as both a bête noire for ID and a much-needed foil. As Dembski wrote, "Thankfully, Richard Dawkins is more explicit than most of his colleagues in making this point and therefore does us the service of not papering over the contempt with which the Darwinian establishment regards those who question its naturalistic bias."[21] ID proponents got tremendous mileage out of Dawkins's rather unsubtle war against all religious belief; he proved preeminently useful as a convenient archetype of the kind of scientist that most anti-evolutionists were convinced was ubiquitous: a perpetually enraged, anti-theistic crusader on a mission to quash religion once and for all. If all scientists truly believed like Dawkins—even the ones milder in demeanor—then ID could brook no compromise with them. Metaphysical bias made a truce impossible. To highlight what Johnson viewed as a metaphysical predisposition, he often featured a quotation from the evolutionary biologist Richard Lewontin as Exhibit A. In Lewontin's words, the irreconcilable conflict between naturalism and miraculous supernaturalism meant that scientists must "take the side of science in spite of the patent absurdity of some of its constructs . . . because [they] have a prior commitment, a commitment to materialism." Lewontin included himself in this and continued, "It is not that the methods and institutions of science somehow compel us to accept a material explanation of the phenomenal world, but, on the contrary, that we are forced by our *a priori* adherence to material causes. . . . Moreover, that materialism is absolute, for we cannot allow a Divine Foot in the door."[22] Such an admission from an eminent scientist was gold dust for an ID supporter, and one can find this quote in almost every pro-ID book.

Most of the ID movement followed Johnson's lead on this position, which served as the starting point for virtually every one of its advocates. For William Dembski, ID was primarily opposed to "blind evolution, not to evolution simpliciter." After all, "materialism is not a neutral, value-free, minimalist position from which to pursue inquiry. Rather, it is itself an ideology with an agenda."[23] Michael Behe, in his later-edition foreword to *Darwin on Trial*, promised readers of Johnson that they would reach an epiphany, and "the skeptic of Darwinism comes to the conclusion that a large part of the modern worldview is built not on solid scientific evidence but on a philosophical bias enforced with sociological prejudice."[24] David Berlinski took Johnson's side in *Commentary*. "The theory of evolution is a materialistic theory," he stated. "Various deities need not apply. Any form of mind is out."[25] For David Keller, "the main contribution Johnson has made to science was in pointing out that the gatekeepers of science hold religiously to their faith in materialism" and that "most scientists do not take time to understand their core assumptions."[26] Even a few antagonists granted the point. John F. Haught, a Catholic theologian and ID critic, hesitantly admitted, "If there is any value in Johnson's work, perhaps it consists in its oblique rebuke to those scientists who recklessly engage in a mixing of science with materialist ideology and then call this amalgam 'science.'"[27]

When it came to the conflict between naturalism and supernaturalism, most of ID's critics took up the same charge in response. Johnson and the intelligent design supporters, they argued, conflated two kinds of naturalisms: methodological and metaphysical. The first was a requirement for science but was not indicative of whether a deity existed; the latter was granted as a philosophical view that could not be directly derived from nature. For Robert Pennock, "The methodological naturalist does not make a commitment directly to a picture of what exists in the world, but rather to a set of methods as a reliable way to find out about the world." It was not a postulate about whether supernatural entities did or did not exist. Pennock roundly criticized Johnson for basing his entire argument around naturalism but providing "only a cursory discussion of the concept."[28] The theologian Keith Ward contended that Johnson put his wedge in the wrong place and should have thrust it between methodological and metaphysical naturalism, rather than between naturalism and theism.[29] ID proponents typically replied that this was a distinction without a difference, as Thomas Woodward did when he summarized the ID position: "the distinction between methodological and metaphysical naturalism is functionally meaningless and misleading, since to exclude intelligent causes from consideration in science is really the same as excluding them from reality."[30] (Such a contention strikes a rather scientistic tone: both IDers and their atheist opponents seem to agree that all "reality" must be scientifically documentable). Dembski likewise wrote that "methodological naturalism is the functional equivalent of a full-blown metaphysical naturalism" and that the solution is not to embrace it but to "dump" it.[31]

The role of metaphysics—and the way it often lurks without acknowledgment in the shadowy umbra behind one's scientific principles—has long been a problem for the philosophy of science. Whereas early creationists like William Jennings Bryan relied on Baconian conceptions of science, later flood geologists warmed to the writings of the philosopher of science Karl Popper (1902–1994) and the historian Thomas Kuhn (1922–1996). Because the courts had fallen back on the demarcation problem (the question of what is or is not science) to bar creationism from the science classroom (since it was not science), creationists started to counter this move by drawing on the philosophy and history of science to argue that Darwinism was not science either, turning the demarcation problem around against their opponents.[32] ID's leaders took up this baton. In *Darwin on Trial*, for instance, Johnson relied heavily on Popper, the august Austrian philosopher proving a useful ally since he had at one point referred to evolution as "unfalsifiable."[33] Johnson also advocated for the epistemic humility of philosophy's rabble-rousing resident bad boy, Paul Feyerabend (1924–1994), who argued that all science had metaphysical backing and that one must recognize this to be, in the words of a famous essay of his, "a good empiricist."[34] As Feyerabend wrote, "A science that is free from metaphysics is on the best way to becoming a dogmatic metaphysical system."[35] These figures aside, probably no figure in the history or philosophy of science exerted a hold over ID as much as Thomas Kuhn, whose 1962 book *The Structure of Scientific Revolutions* supplied for ID a blueprint on how to topple and replace the Darwinian worldview.

ID's affection for Kuhn went back at least to Denton, who deployed him at great length. In Kuhn's understanding of history (he wrote *Struc-*

ture as a historian but later addressed more philosophical issues), science was governed by successive paradigms, and the process by which one was replaced by another was more haphazard than is often realized. "Normal science" was the general stage under which most scientific work was done, but over time, each paradigm (say, for instance, Ptolemaic cosmology or Newtonian mechanics) would start accruing more and more isolated irregularities, what Kuhn termed "anomalies." Eventually, these anomalies become too numerous and prominent to ignore, and the paradigm enters a period of "crisis," after which a new paradigm arises to take its place. His phrase for this transmutation—"paradigm shift"—has entered the vernacular.[36] The details of Kuhn's history have been controversial—widely hailed and widely disputed—but his concept of the paradigm shift was central to ID's self-understanding and hope for the future.[37]

We have already seen, in Marsden's writing, how the paradigm shift from Baconian to theoretical science bedeviled earlier fundamentalists.[38] But, for ID, Kuhn's theory of history would prove an ally, and paradigm shifts were welcomed, not spurned. Michael Denton took up Kuhn's vision of the history of science in *Evolution: A Theory in Crisis*, and the subtitle of his first book indicated his belief that evolutionary theory was in a crisis mode in the same manner as Ptolemaic cosmology was in the early modern period. In his final chapter, with the Kuhnian title "The Priority of the Paradigm," Denton alleged that the anomalies were approaching critical mass, writing that some aspects of the modern synthesis appeared "strikingly reminiscent of the mental gymnastics of the phlogiston chemists or the medieval astronomers." However, the Darwinian establishment ignored or downplayed the force of the anomalies because, as Denton saw it, any alternatives were simply inconceivable (or perhaps emotionally distasteful). "The history of science," wrote Denton, "amply testifies to what Kuhn has termed the 'priority of the paradigm,' and provides many fascinating examples of the extraordinary lengths to which members of the scientific community will go to defend a theory just as long as it holds sufficient intrinsic appeal." The upshot was that the devotion to a paradigm meant that dissenters would be met with indignation and incomprehension. Whether Kuhn's view of history was right or not, continued Denton, "it certainly provides a satisfying explanation of why even in the face of what are 'disproofs,' Darwinian concepts continue to dominate so much of biological thought today."[39]

Following Denton, Johnson drew on Kuhn much to the same effect in *Darwin on Trial*, taking up a deconstructionist approach to the history of science and contending that Kuhn demonstrated that paradigms were "a way of looking at the world that is influenced by cultural prejudice as well as by scientific observation" and that "a paradigm is not merely a hypothesis, which can be discarded if it fails a single experimental test." As a result, those scientists trained and operating under the normal science of a paradigm were often unable even to see the problems that would be clear to an outsider (perhaps a lawyer). "Kuhn described experimental evidence," wrote Johnson, "showing that ordinary people tend to see what they have been trained to see, and fail to see what they know ought not to be present. The finest scientists are no exception." Kuhn might have focused on astronomy, but Johnson wanted to push him to biology, too, and he surmised, "If Kuhn had chosen evolutionary biology as a case study, he would have risked being denounced as a creationist."[40]

For some ID advocates, the resistance they experienced from mainstream scientists was further evidence that confirmed Kuhn's view of history. The ID supporter and literature scholar Jonathan Witt summarized this perspective neatly: "Kuhn . . . highlighted what is today a truism among historians of science: Reigning scientific paradigms, challenged by new and contrary evidence, do not go gently. Most scientists who invest their careers in a paradigm find it hard to admit the mistake."[41] Jonathan Wells (1942–2024) attributed design's lack of acceptance to the difficulty that new ideas face in gaining acceptance from the established scientific community, something he believed Kuhn demonstrated. In an interview with Wells, Casey Luskin raised an oft-quoted line from Kuhn's work. "No part of the aim of normal science is to call forth new sorts of phenomena," wrote Kuhn; "indeed, those that will not fit the box are often not seen at all. Nor do scientists normally aim to invent new theories, and they are often intolerant of those invented by others." Upon hearing this, Wells stated that he agreed with the sentiment and that it helped explain the resistance design proponents faced from scientists. Disputes between older and newer paradigms could be vicious no matter the field, he reasoned, but there was more at stake with Darwin and design, because naturalism itself was the concept under review. "This particular clash is sharper and more fundamental than some other disputes in science," he elaborated.[42] In the same vein, a group of ID supporters (in responding to the *Dover* ruling) reiterated the same Kuhn quotation and argued, "New ideas are often bitterly opposed by the champions of the existing orthodoxy in science, and mainstream science journals frequently refuse to publish outside of the existing paradigm." Furthermore, "scientists with 'productive careers' may exhibit '[l]ifelong resistance' to new paradigms in science."[43]

Kuhn gave ID advocates an explanation for why their theories should be met with such vigorous resistance, but he also gave them a big-picture framework for how the struggle should play out. "Now I find it interesting," Wells continued in his interview with Luskin, "that Kuhn describes certain features of scientific revolutions that we actually see in this present controversy. One of them is, in the process of a scientific revolution, the very meaning of science—the definition of science—comes into dispute. And we see that happening here. A lot of the clashes we read about are arguments over what is science."[44] Science's definition, however, did not necessarily reduce to simple empiricism. In a later work, Wells focused on the social and political context of the paradigms, writing, "If paradigms are essentially circular (Kuhn called them 'incommensurable'), then the choice between them depends not on logic and evidence but on irrational psychology and power politics."[45] The potential for relativism, and even irrationality, in Kuhn's work has been debated by scholars ever since he published his watershed work. The philosopher of science Peter Godfrey-Smith, for instance, lamented the existence of the tenth chapter of Kuhn's book, in which he became more adventurous and seemingly suggested that paradigms can in some sense determine reality. Calling it "X-Rated Chapter X," Godfrey-Smith wrote that it would have been better if Kuhn had "left this chapter in a taxi, in one of those famous mistakes that authors are prone to." He concluded, however, that Kuhn was not ultimately a relativist on the most important issue—that of scientific progress.[46] Robert Pennock found ID's use of Kuhn to be more or less postmodern, writing that ID's "interpretation of Kuhn is a red flag that should alert us to look carefully at the notion of scientific truth that they have in mind."[47] In a later essay, Wells distanced himself (and ID) from the relativistic interpretation of Kuhn, writing that Kuhn's real problem was that he viewed science in a Darwinian fashion. Progress in scientific knowledge was only possible because it is analogous to evolutionary development, and the best scientific theories are selected for their survival value. In deploying Darwinian language, Kuhn argued, "I am a convinced believer in scientific progress." But for Wells, this was further evidence that evolutionary theory had no real concept of truth under it. "It seems," concluded Wells, "that even Kuhn admitted that unguided processes do not solve problems or lead to truth; intelligent direction is necessary."[48] Kuhn was therefore useful not only in showing that Darwinism could be treated as a paradigm in need of replacement but also in showing the relativistic notions of truth that are implied by the Darwinian conception of reality. In drawing on Kuhn, postmodern or not, ID advocates found a way to understand their struggle. A Kuhnian contest was an opportunity: a paradigm crisis could open the way for ID to displace naturalism. Wells, again quoting Kuhn, noted that if the promoters of a new paradigm (a non-Darwinian one) were "competent, they will improve it, and explore its possibilities, and show what it would be like to belong to the community guided by it." This scientific revolution, for Wells, "has been happening with ID."[49]

Many of ID's leaders suggested that design must become the alternative paradigm to replace Darwinism. The Discovery Institute president Bruce Chapman related in the postscript to *Mere Creation* that ID was challenging materialism, such that it would "no longer be assumed unquestioningly."[50] Likewise, Wells argued, "the new paradigm . . . will be based on design."[51] But Dembski acknowledged Kuhn's declaration that, as he paraphrased, "you can't shift into a vacuum."[52] As Kuhn wrote, and Denton quoted, "the decision to reject one paradigm is always simultaneously the decision to accept another."[53] Highlighting supposed anomalies in evolution was not enough. Denton, of course, predated the ID movement's rise with *Evolution: A Theory in Crisis*, and so he supplied no alternative. Later ID advocates would take up the cause. In Wells's words, "for a design paradigm to out-compete Darwinism, however, it will have to be developed to the point where it is philosophically rigorous and scientifically fruitful."[54] That said, design could only be embraced once methodological naturalism had been abandoned and metaphysics was allowed back into the picture. This put ID in a rather precarious position. How could one demonstrate that the naturalist paradigm was breaking down and that ID was ready to take its place, if, in their view, any alternative to Darwin was unthinkable? Kuhn supplied the historical template: overload the current paradigm by stressing its anomalous characteristics until it was on the brink of collapse and then build up ID enough so that it could function as a scientifically and philosophically licit substitution.

Practically, this meant that most of the ID attacks on Darwin were focused on offense, not defense. Major ID proponents used science in a rather instrumental way as a result, often less interested in proving design than in proving the insufficiency of Darwinism. As Johnson wrote, "I felt no obligation to offer my own theory about how life was created in the first place. . . . My purpose was to show that what is presented to the public . . . is mostly philosophical speculation."[55] In fact, Johnson only used the term "intelligent design" a handful of times in *Darwin on Trial*, and even then usually as a general term for creation. He did, however, refer to intelligent design as a "paradigm" in the book's research notes.[56] It was only after *Darwin on Trial* that the movement of intelligent design began to coalesce, with some of Johnson's followers building both a scientific critique and an eventual replacement. The most notable of these figures was Dembski, who felt that the Kuhnian shift could only take place once design had enough of a foundation to stand on its own. Then, once the paradigm collapsed, design could take its place—a paradigm hospitable both to theistic metaphysics and, by eventual extension, to social conservatism. This, then, was the focus of the science of ID.

The Science of Intelligent Design: Stressing the Anomalies, Building the Paradigm

The two major scientific arguments that buttressed the ID movement in the decade-long aftermath of *Darwin on Trial* were Michael Behe's irreducible complexity and William Dembski's specified complexity. Following on Johnson's heels, Behe and Dembski took up the mantle of scientific and philosophical criticism to help overload the evolutionary paradigm and pave the way for a pivot to design. They both prosecuted a negative attack, spending much of their time rebutting and refuting Darwinism and comparatively less time offering an alternative (though, eventually, Dembski did go further and begin to build a design-centered replacement). There were, of course, a great deal many more ID arguments made by other writers (and it must be noted that they did not originate these concepts either, as both were described by Norman Geisler in *Origin Science*: specified complexity by name, and irreducible complexity by biochemical concept). But these two prongs garnered the most media attention and critical response, and it was on them that the ID enterprise depended, because unless the Darwinian paradigm could be undone, then design would be stillborn.

Irreducible Complexity

Behe boasted scientific qualifications that few in the world of anti-evolution could rival. A practicing Roman Catholic with a PhD in biochemistry from the University of Pennsylvania, Behe has spent most of his career at Lehigh University as a professor of biochemistry. As a Catholic, Behe had little background in the fundamentalist world of anti-evolution, and so he recounted that he was a theistic evolutionist for most of his youth. While at Lehigh, however, he encountered the work of both Denton and later Johnson, and the Berkeley lawyer "got [him] moving." When the philosopher of science David Hull (1935–2010) published a brutal review of *Darwin on Trial* in *Nature*, Behe sent a self-described "really swell letter" to the editor in Johnson's defense, but it went unpublished. Behe was frustrated by the association of Johnson with Henry Morris's Institute for Creation Research, a tarring and feathering that he felt implied "guilt by association." John Buell, at the Foundation for Thought and Ethics (the same think tank that had sponsored *The Mystery of Life's Origin* and *Of Pandas and People* a decade earlier), later invited Behe to participate in a conference on evolution. In meeting and mingling with the leading lights of the new anti-evolution, Behe began to formulate some ideas on how to take the war on Darwin to a new front. Like Johnson before him, and following the common populist strain of anti-evolution, Behe decided that writing a trade book on the issue would be the best strategy, and the result was the most important and controversial work on intelligent design up to that point: *Darwin's Black Box*.[57]

Published in 1996 by the Free Press, *Darwin's Black Box* made waves owing to the credentials of its author and the prominence of its publishing house (indeed, the Free Press was a far cry from the shabby, self-published books of the scientific creationists of yesteryear). *Black Box* also impressed its sympathetic readers with its sense of scholarly rigor, especially those readers untrained in the sciences. Warnings about the density of the text affixed the headiest sections on biochemical systems, and the perplexed were invited to skim the portions if needed.[58]

For the overall attack, Behe focused on one of Darwin's own quotations: "If it could be demonstrated that any complex organ existed which could not possibly have been formed by numerous, successive, slight modifications, my theory would absolutely break down."[59] Behe contended that discoveries in biochemistry "pushed [Darwinian evolution] to its limits."[60] Indeed, biochemical systems were so multifaceted and complicated that it was impossible, in Behe's view, for them to have arisen in the piecemeal way that Darwin envisioned. The bacterial flagellum motor became the poster child for Behe's argument, but he also documented the complexity of cellular cilium, the protein cascade that causes blood clotting, and the old design standby of the eye, among other things. Behe regarded these biochemical mechanisms as "irreducibly complex"—that is, they were "composed of several well-matched, interacting parts that contribute to the basic function, wherein the removal of any one of the parts causes the system to effectively cease functioning." Such a system could not be produced by "slight, successive modifications of a precursor system."[61] In a bold chapter titled "Publish or Perish," Behe surveyed the literature on biochemical evolution and declared that there was not one peer-reviewed paper explaining how such irreducibly complex systems could have evolved. For Behe, these lacunae were proof of a great crisis in evolutionary theory; if a theory claimed to illustrate some phenomenon but offered no "attempt at an explanation, then it should be banished."[62]

Criticism was immediate and wide-ranging. The evolutionary reply often took the tone of Mark Twain's exasperated traveling companion in *Roughing It*, who—when his story of being chased up a tree by a climbing bison was doubted by his companions—exclaimed, "Because you never saw a thing done, is that any reason why it can't be done?"[63] Jerry Coyne argued that just because one could not envision how a complicated system evolved, "it is not valid . . . to assume that [such pathways] could not have existed."[64] Kenneth Miller, an evolutionary biologist and a Catholic, wondered whether creationists understood how much ground Behe had ceded to evolution, noting that he accepted common descent and an ancient Earth.[65] Miller took great pains to lay out plausible pathways along which such biochemical systems might have evolved and promoted the work of Russell Doolittle (1931–2019)—whose findings Behe disputed—which "has not only shown how such a complex system *might* evolve, but has also produced comparative studies showing how it probably *did* evolve."[66] To declare otherwise was to commit that great philosophical error that has pinballed through the history of religion and science, the dreaded "God of the Gaps"—the charge that, whenever someone did not understand how something happened in nature (when, that is, there was a gap in our scientific knowledge), one should not just chalk it up to a miracle and call it a day. Critics alleged that this was just what Behe had done.

Behe responded by doubling down on irreducible complexity in his decade-later book *The Edge of Evolution* (2007), though he clarified some of the terminology of the argument. In the light of biochemistry, he asked, "Can we determine not what is merely theoretically possible for Darwinian evolution, not what may happen only in some fanciful Just-So story, but rather what is *biologically reasonable* to expect of random mutation and natural selection at the molecular level?" To determine if something was "biologically reasonable," Behe suggested a two-pronged approach: to look at the "intermediate evolutionary steps that must be climbed" and then at the "coherent ordering of steps toward a goal." When looking at evolutionary steps and the coherence of the unfurling plan, which—design or blind evolution—was more reasonable?[67] In *Edge*, Behe clarified that irreducible complexity did not mean that it was logically impossible for a complex biochemical system to have evolved through random mutation and natural selection but rather that it is implausible given the alternative of design. He later hedged this a bit further by writing that irreducible complexity did not mean "complex biochemical systems must have been created *ex nihilo*." No, said Behe, "I have never claimed that. I have no reason to think that a designer could not have used suitably modified pre-existent material." The argument here, then, is that the complexity of the system, even if a pathway toward its development could be conceived, indicated design because it was implausible to suspect that it happened in an unguided manner. "My argument in *Darwin's Black Box*," he continued, "is directed merely towards the conclusion of design. How the design was effected is a separate and much more difficult question to address." Perhaps it could be done evolutionarily through "directed mutations." Highlighting the difficulty in resolving divine action in the world (such as the kind that might guide evolution), Behe contended that Kenneth Miller's belief in God's action through quantum events indicated that he believed in a form of intelligent design.[68]

The perceived shiftiness of Behe's claims frustrated his critics—for if irreducible complexity was merely probabilistic, or possibly even open to evolution, then what good was it? Dembski defended Behe on this point, though he admitted that irreducible complexity needed some "fine-tuning." "Behe isn't saying it's logically impossible for the Darwinian mechanism to attain such systems," Dembski wrote. "It's logically possible for just about anything [to happen]." However, for example, just because it is conceivable for Mount Rushmore to have been caused by erosion would not make such a belief reasonable. For Dembski, the crucial issue—the difference between direct and indirect pathways—needed teasing out of Behe's thinking. One could understand a direct pathway as a system evolved for a consistent purpose, such as the heart to pump blood. An indirect pathway would be for a system to change its function through its evolutionary history, like if the heart began life as a defense mechanism that made throbbing noises to ward off predators and later evolved into the central organ for pumping blood. For Dembski, irreducible complexity is a contention that many "biochemical systems are provably inaccessible to direct Darwinian pathways. . . . The Darwinian mechanism has no intrinsic capacity for generating such systems except as vastly improbable or fortuitous events." Normally, evolutionary biologists speculated that such systems were jury-rigged indirectly, with adaptations that formerly had other purposes coming together to function in a new way, a phenomenon called "exaptation." ID advocates contended that indirect pathways failed too, for, they argued, no such pathways were known for any complex biochemical system. According to Dembski, "What's needed is a seamless Darwinian account that's both detailed and testable of how subsystems undergoing coevolution could gradually transform into an irreducibly complex system. No such accounts are available or forthcoming." This did not reduce Behe's argumentative power, however; Dembski contended that irreducible complexity did not mean "that certain biological systems are so complex that we can't imagine how they evolved by Darwinian pathways"—the reading that critics took of Behe—but rather that "we can show conclusively that they could not have evolved by direct Darwinian pathways and that indirect Darwinian pathways, which have always been on much less stable ground, are utterly without empirical support." Though Behe attacked Darwin at length, he only obliquely referenced an alternative: intelligence. Intelligence was the only known cause capable of bringing about irreducible complexity, and so design must be considered as an alternative hypothesis.[69]

For all the clarity of Behe's prose, his conception of design was ambiguous, and this ambiguity posed problems for both friends and opponents. In his brisk treatment of ID in *Darwin's Black Box*, Behe defined design as the "purposeful arrangement of parts"—a definition too minimal to satisfy the critics who wanted more precision. When drawing demarcations on designed systems versus naturally evolved ones, Behe timidly observed, "Some features of the cell appear to be the result of simple natural processes, others probably so. Still other features were almost certainly designed." It did not help that Behe later contended that the main reason scientists could not see design in nature was because of their aversion to God.[70] Criticism came from every direction. Evolutionists such as Jerry Coyne called his argument creationist; creationists such as Norman Geisler called him an evolutionist.[71] Coyne described the alternative of design as a "confusing and untestable farrago of contradictory ideas."[72] Geisler lamented Behe's agreement "with the [evolutionary] party-line criticism that this positing a series of creative events after the beginning is an unnecessary God-of-the-gaps move."[73] The leveling of such contradictory charges indicated a lack of cohesion in the advancement of design. Behe's argument was almost entirely negative; it contained little on which one could build an alternate worldview. It would fall to William Dembski to articulate a fuller vision of what design meant as a scientific concept and to move ID beyond a one-sided attack on Darwin so that it might become an actual paradigm that could replace it. Though Behe's description of design was minimal, however, his concept of irreducible complexity would remain a mainstay of design theory and featured prominently in the 2005 *Dover* trial that later proved a watershed moment in the history of the movement.

Specified Complexity

William Dembski specified the other complex argument of note. Dembski came from a family without qualms about Darwin; his father had a doctor of science from Germany in biology and taught evolution in college.[74] After undergrad study at the University of Chicago, the young Dembski went on to earn two doctorates, one in mathematics from his alma mater and the other in philosophy from the University of Illinois at Chicago—and then added a master of divinity from Princeton Theological Seminary. He was a postdoctoral fellow at Northwestern under the Darwinian philosopher of science David Hull.[75] His religious background also changed throughout the years. Dembski chronicled that he did not have any religious convictions while younger but later became a Christian, specifically a Baptist. In the late '90s, Dembski converted to Eastern Orthodoxy, though he returned to a Baptist denomination a few years later.[76] His broad expertise impressed his colleagues and frustrated his opponents. Phillip Johnson referred to him as "'Bill of the bulging brain and bulging muscles' because he works out and it shows," while Barbara Forrest wrote that his obvious gifts made him "perhaps the saddest of the ID *personal* stories, for Dembski began his crusade . . . with intelligence and scholarly promise."[77]

Dembski laid out his argument regarding specified complexity in his 1998 Cambridge-published *The Design Inference*, which was the first book of the ID movement to be released by a peer-reviewed publisher. Dembski sought to frame the design inference as a logical process, resulting from a piecemeal elimination of the alternatives of chance and necessity through small probabilities. This inference could be made using the "explanatory filter." "Whenever explaining an event," Dembski wrote, "we must choose from three competing modes of explanation. These are *regularity*, *chance*, and *design*."[78] To arrive at design at the end of the filter required a different method than the probabilistic elimination that might take place in statistics. A "design theorist," rather, "is in the business of categorically eliminating chance."[79] Design was a conclusion to be reached cautiously and after one had exhausted other options, found only after eliminating chance and regularity.

Information formed the bedrock of Dembski's arguments. A random sequence of letters in alphabet soup (for instance, sghauhiuwhdoasf-

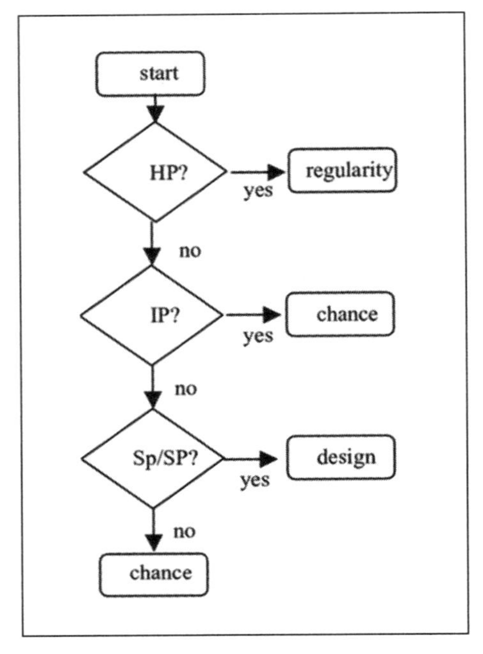

Figure 2.1. The explanatory filter, running through the gauntlet of high probability (HP), intermediate probability (IP), and small probability (SP). See Dembski, *Design Inference*, 37.

haiusgbh) may be a highly improbable, complex string of letters, but it contains no informational content. However, if the soup contained a sentence with specific informational content ("look out behind you"), one would recognize that this is a pattern attempting to communicate (and, probably, one would be alarmed). Such a message would not be attributed chance; it would be both complex and specified.

What about law? In Tom Stoppard's play *Rosencrantz and Guildenstern Are Dead*, Rosencrantz examines a flipped coin that comes up heads eighty-five times in a row. Guildenstern is "worried by the implications" and observes that "a weaker man might be moved to re-examine his faith, if in nothing else at least in the law of probability." Because the odds of such a sequence virtually preclude it as a possibility, perhaps the only plausible answer would be that it was metaphysically necessary for the coin to come up heads so many times in a row. But there is no known metaphysical law governing coin flipping, and the closest similar concept, the "law of probability," states that anything with a 50/50 possibility (such as a coin flip) should be roughly equal when tallied up at the end of so many flips. If Dembski were sitting on the sidelines watching Rosencrantz and Guildenstern, he would suggest a different approach. The string of heads resulted not from chance or law, so only one alternative remained: design. Someone, an intelligent creator (Tom Stoppard, perhaps?), had designed it thus.[80]

As a case study, Dembski turned to the "creation-evolution controversy." The argument for the design inference regarding the origin of life assumed this form:

Premise 1: LIFE has occurred.
Premise 2: LIFE is specified.
Premise 3: If LIFE is due to chance, then LIFE has small probability.
Premise 4: Specified events of small probability do not occur by chance.
Premise 5: LIFE is not due to a regularity.
Premise 6: LIFE is due to regularity, chance, or design.
Conclusion: LIFE is due to design.

Dembski analyzed the responses to this argument from Richard Dawkins (chance), Stuart Kauffman (regularity), and Frank Tipler and John Barrow (chance). All these responses tried to block premises 3 and 5, though with different reasoning. Dawkins argued, along the lines of the Miller-Urey experiment, that a chance combination of chemicals sufficed. Kauffman contended that life resulted from the spontaneous organization of order, the self-organization thesis, which Dembski characterized as law/regularity. Lastly, Barrow and Tipler took up the multiverse interpretation of quantum mechanics, which meant that anything of small probability could be realized somewhere. Dembski rejected these various alternatives to design. In his view, they failed to pass the explanatory filter.[81]

As an alternative, design often seemed like a negative conclusion. "To infer design is to eliminate regularity and chance," Dembski summarized. Could simply eliminating these two things form the basis of another option? Addressing this question, Dembski tried to expand the concept and show that the articulation of design could boast positive content. Inferring design was more than just canceling out regularity or chance, he amended, but also included attempts "to detect the activity of an intelligent agent." The explanator filter "coincides with how we recognize intelligent agency generally," and design fit the bill as a framework that relied on neither chance nor regularity.[82] Despite this, for Dembski, even though the filter could potentially do more than attack the reigning paradigm—it could theoretically identify agency—the theological or philosophical characteristics of such a designer were beyond the bounds of science to discover. The filter may be able to declare that some designing agent existed but could not tell us anything about the nature of that agent.

Another example can further clarify Dembski's position. He was fond of turning to SETI (the Search for Extraterrestrial Intelligence) as proof that the design inference was already an acceptable part of scientific research (provided it had nothing to do with God). Dembski cited the movie *Contact* (based on the Carl Sagan novel) as a paradigmatic explanation of the kind of investigation a SETI program would undertake. As he wrote in *Intelligent Design*, the movie's SETI researchers looked for radio waves as evidence of intelligence—but, complicating matters considerably, "many natural objects in space produce radio waves (e.g., pulsars)." Trying to sift through all that noise would be like "looking for a needle in a haystack." To make it easier, SETI researchers try to find patterns in the noise. In *Contact*, irrefutable evidence of intelligence comes when a long binary message is received (communicated via beats and pauses) that represents the prime numbers 2 to 101. This is seen as "decisive confirmation of an extraterrestrial intelligence."[83] Why? Because it addresses three criteria for distinguishing design from nondesign: contingency, complexity, and specification. He writes, "Contingency ensures that the object in question is not the result of an automatic and therefore unintelligent process that had no choice in its production. Complexity ensures that the object is not so simple that it can readily be explained by chance. Finally, specification ensures that the object exhibits the type of pattern characteristic of intelligence."[84] A real-world example would be Breakthrough Listen, which began in 2016. Breakthrough Listen searches for radio waves to try to find a "signal that bears the hallmarks of a technological society." For example, "Alien civilisations could be leaking radio emissions, in the same way TV broadcasts and radar signals on Earth spread out into space. Or they could be transmitting greetings into the space [*sic*] in the hope that someone is listening."[85] For Dembski, this kind of research showed that detection of intelligence was already a part of science.

Dembski was adamant that though ID had a long way to go, it bore all the hallmarks of a paradigm that would not simply delete an established theory in Darwinism but rather add to biology and become itself a testable research program. Where Johnson and Behe argued primarily negatively, Dembski acknowledged the need to go further. Once ID laid out the criteria for detecting intelligence, applied those criteria to biological systems, and then showed that those criteria ruled out material mechanisms, ID would be well on its way. He argued that, by the year 2000, these steps would have been taken, and so at the turn of the millennium, ID would reconceptualize biology by adding a design-centric framework to scientific investigation. This development would eventually inspire a program of biological research within that paradigm. He admitted that some of ID's program was still a "promissory note" but expressed confidence that ID would achieve its task.[86] Because intelligence was a known cause for information that was specified and complex, it was reasonable to begin investigating whether intelligent causes could act, sculpt, and create in nature. Unguided evolution provided only "handwaving just-so stories," but intelligence was recognized by all as a valid designing agent in examples of engineering—so why should it be excluded while "we're . . . waiting for the promised material mechanisms?"[87] ID could, he argued, assist science. As an example, Dembski highlighted the shift in understanding "junk DNA" (that is, noncoding DNA). Scientists began to move away from the idea that most of the human genome was "junk . . . due to the sloppiness of the evolutionary process" to a more open stance regarding noncoding DNA's involvement in epigenetic activity or with gene regulatory networks. For Dembski, design theorists needed to be at the forefront of this research, for a design paradigm would have predicted these findings.[88]

The abstract and highly technical nature of *The Design Inference* made Dembski's argument one of the least accessible contributions to intelligent design, even though it carried a great deal of strategic weight. Dembski attempted to rectify this with a popular version of the same argument in his book *No Free Lunch*; but even that version was more difficult to grasp than Behe's (or, later, Stephen Meyer's argument about the complexity of the cell), and it did not factor into the 2005 *Dover* case with the ubiquity that irreducible complexity did.[89] However, these two arguments formed the backbone of ID's initial campaign against evolution, and they supplied anti-evolutionists with ammunition the likes of which they had arguably not seen since before the emergence of the modern synthesis (and where they did draw on—or at least bear similarity to—earlier creationist arguments, the ID versions were better packaged and presented).

After all this, though, a major question remained. Suppose (granting ID the argument) the design paradigm was coming into its own: What did that mean for the identity of the designer? Who was it? Was it God? Could the designer's identity be known? ID proponents may have accused evolutionists of metaphysical bias, and they may have built up scientific concepts designed to erode confidence in unguided evolution, but how did their own metaphysics influence their work? What was their theology? Did they have one?

The Theology of Intelligent Design: William Paley and Natural Theology

After Darwin, the man who figured most prominently in discourse about design was William Paley (1743–1805), an eighteenth-century Anglican clergyman whose most famous work was the aptly titled *Natural Theology*, published near the end of his life in 1802. Paley opened this work with the famed watchmaker analogy. Imagine, he wrote, that one happened upon a watch lying in a ditch. Surely, such artificial machinery must have had a designer. He wrote, "This mechanism being observed (it requires indeed an examination of the instrument, and perhaps some previous knowledge of the subject, to perceive and understand it; but being once, as we have said, observed and understood), the inference, we think, is inevitable, that the watch must have had a maker: that there must have existed, at some time and at some place or other, an artificer or artificers who formed it for the purpose which we find it actually to answer, who comprehended its construction and designed its use."[90] As with a watch, he continued, so with biological organisms, whose artifices were machine-like in their breathtaking complexity. The eye, for instance, was designed with such wondrous purpose that it all but cried out in testimony of its divine designer. The eye was like a telescope, and such a complicated piece of machinery could not have arisen without a designer making it.[91] In this brief passage, one finds the dual emphasis of Paley's study: an interest not only in adaptations but also in complexity.

ID's critics were quick to point out the similarities that it bore to old versions of the design argument, Paley's in particular. "Essentially," wrote the philosopher of science Michael Ruse, "[intelligent design] is the argument of Plato, of Aquinas, and of Paley. There is nothing to be ashamed of in this, but there is nothing to be excited about this either."[92] (Thomists would later dispute that ID's design arguments were similar to any of Aquinas's "Five Ways.") The biologist Bruce Grant castigated ID proponents as "Neo-creationists [who] imitate Paley's designed-watch metaphor and peddle it like a Hong Kong Rolex."[93] Robert Pennock argued that Stephen Meyer paid "homage to the classic design argument from William Paley's *Natural Theology*" but updated it to feature new kinds of complexity (rather than Paley's older examples, which strike today's readers as "ridiculous").[94] Ronald Numbers wrote that *Darwin's Black Box* cemented Behe as "a modern-day William Paley."[95]

When ID advocates were accused of clandestine creationism by their opponents, they typically responded with furious denials, but when connected to Paley, they were rather agreeable. The ID historian Thomas Woodward noted that ID proponents hoped to "resurrect" Paley's main idea, "the *inference to design*."[96] Similarly, Michael Denton argued that Paley was correct to use the mechanical analogy in understanding nature and, in turn, to use that as an argument for design. "Paley was not only right in asserting the existence of an analogy between life and machines," he wrote, "but was also remarkably prophetic in guessing that the technological ingenuity realized in living systems is vastly in excess of anything yet accomplished by man."[97] Following Denton, Michael Behe made probably the most extensive use of Paley, defending him at length in *Darwin's Black Box*. Just as Paley did, Behe embraced the machine analogy to nature, writing in a later work that Paley adumbrated the "conceptual difficulty" at the heart of Darwinian theory, that "the cell too is chock-full of elegant machinery that requires multiple interacting components."[98] Behe devoted time to the complexity of the eye as an example of irreducible complexity (much as Paley did, though without the same terminology) and noted that Darwin himself was deeply troubled by the eye. Behe boldly stated that it was "surprising but true that the main argument of the discredited Paley has actually never been refuted." Paley might have overreached in some of conclusions (perhaps in his speculation about God's nature), but the later refutation of his excesses unfairly became "a refutation of Paley's main point, even by those who know better."[99] Woodward encapsulated the entire strategy succinctly when he wrote that ID proponents wanted "to fully vindicate the story of Paley."[100] ID redeployed Paley's argument as a biological problem in need of solving, and they embraced machinery as an appropriate metaphor for nature.

The machine analogy is key not only to intelligent design but also to the design argument as a whole. The molecular biologist Douglas Axe, for instance, wrote that technological development provided the most intuitive way to understand design: "technology brought ever more compelling comparisons between the handiwork of humans and the handiwork of God—from Plutarch's shoe, to Paley's watch, to Behe's mousetraps, to Meyer's computer." Just as such complex devices were obviously designed by human intelligence, so the obviously complex world-machine was designed by a supernatural intelligence.[101] The inference to the best explanation hinges on equating major features of the universe—cosmic fine-tuning, DNA, cell structures—to machinery. The analogy is important because it accords with our day-to-day lives. As Stephen Meyer wrote, "In our experience, intelligent agents often produce finely tuned machines, systems, or strategies in order to achieve discernible functional outcomes."[102] Some ID advocates would go even further and argue that this equation was not simply an analogy but was a literal statement of fact. "Is the inference to design in biology based on a mere *analogy* to other forms of design, by humans?" asked Casey Luskin on the *Evolution News* flagship website. No, he contended, it is a statement of being, not metaphor: "When we look at the mathematics and physics behind what is going on, the similarities between biological machines and biological information and written language and human-designed machines go beyond mere analogy and become an identity."[103]

Michael Behe is possibly the most adamant of all ID thinkers regarding the use of machine language. In the years after *Darwin's Black Box*, Behe insisted that the flagellum motor, one of the main pieces of biochemical evidence in his argument, "is not *like* a machine, it is a *real* molecular machine."[104] When criticized by some of his Lehigh colleagues, Gregory Lang and Amber Rice, for overrelying on the machine analogy, falsely equating proteins to machinery, Behe contended that he was not using an analogy at all. Proteins are "literal machines," he replied: "'molecular machine' is no metaphor; it is an accurate description. Unless Lang and Rice are arguing obliquely for some sort of vitalism—where the matter of life is somehow different from nonliving matter—then of course proteins and systems such as the bacterial flagellum are machinery. What else could they be?"[105] This is central to Behe's argument against Darwin. The rise of biochemistry as a discipline pulled back the veil on the molecular life and so—for Behe—showed that while Darwin appeared to have been on to something, he was further off the mark than most people realized. Wrote Behe, "The startling challenge that modern working science threw down for Darwin's hypothesis was the discovery of *machines* and *information* at the foundation of life. Quite literally, life—like the Borg in *Star Trek*—depends on tiny machines and computers made of molecules."[106] Paley was vindicated because of the discovery of the world's mechanical nature—an intricately designed machine just like that watch he found on his evening stroll.[107]

Not every ID advocate has warmed to the machine analogy—or identity—with such gusto, however. Michael Denton, though initially predisposed to it, eventually turned against the "superwatch" conception of nature. He explained this change in a later essay, writing that he had incorporated the understanding of nature as mechanism into his work when he wrote *Evolution: A Theory in Crisis*, but by the time he had moved on to a later book—*Nature's Destiny*—he had moved away from this perspective. "*Nature's Destiny*," he wrote, "represents my state of mind in transition from a mechanistic to a lawful and naturalistic conception of life."[108] Denton's intellectual journey took him away from figures like Behe, but some other ID supporters have made similar observations. As crucial a figure as William Dembski has expressed similar reservations about the overuse of mechanical language, writing many years later that that even though ID used such arguments to refute Darwin, "that does not mean intelligent design is committed to a mechanistic or reductionist or artifactual view of life or the universe." Debating Darwin means, for a time, looking at biochemical life in mechanistic terms, but once materialism has been defeated, then such language can be shelved. "It is therefore mistaken," he contended, "to think that intelligent design looks at life in the same way as a materialist biology, only that intelligent design sees the hand of a supernatural external designer where a materialist biology sees the hand of natural selection." Furthermore, "to say that organisms exhibit artifactual features is not to say that they are artifacts, any more than saying because humans can perform computations, therefore they are computers."[109]

ID's foes, both theistic and atheistic, have not been so easily mollified. Going back to even before Paley, David Hume (1711–1776) opined that the analogous equation of the universe to a machine was simply a hermeneutical decision, not something self-evident in nature. Philo summarizes Cleanthes's design argument as "the world . . . resembles a machine, therefore it is a machine, therefore it arose from design." But is not this analogy the same as saying, "The world . . . resembles an animal, therefore it is an animal, therefore it arose from generation"? Why not take an organic approach? Maybe the universe is a big vegetable.[110] The historian of technology Lewis Mumford likewise vituperated against anyone using machine analogies for nature, writing that such an idea was part and parcel of Cartesian modernity. Making the world a machine necessitated a reductionism that reversed the developmental process of life. "The machine model," Mumford wrote, "introduced teleology or finalism in its classic form: a purposeful organization for a strictly predetermined end" (note the similarity in definition to Behe's). This was exactly the wrong thing to do, for Mumford, because it "corresponds to nothing whatever in organic evolution." As such, the machine model screams for design, but it has created the very problem that design seeks to fix. Attempts to detect design in the machine world actually militate against organic life as it is (a perpetual preoccupation for Mumford), and he dourly suggested that "the cult of anti-life secretly begins at this point, with its readiness to extirpate organisms and contract human wants and desires in order to conform to the machine."[111] Darwin, contended Mumford, actually struck back against the machine, showing how organic the world actually was by proposing a way for common descent—real organic unity, an anti-mechanical idea—to have occurred. Darwin could have been remembered as a great ecologist, but unfortunately—for Mumford—he gave unwitting license to other expressions that flow from mechanical philosophy: particularly capitalism.[112] It is important to dwell on this because many of ID's critics—both theist and atheist—see it as a central irony to the movement. Despite decrying modernity and its seeming secularism, so goes this riposte, ID has unwittingly embraced a highly mechanical and modernist conception of reality—allying itself with the very reductionism it claims to be resisting and postulating a modern conception of divine action in the world.

As William Dembski asked in *Being as Communion*, where is the room in nature for special divine action (that is, specific instances of God's intervention into the world)? Where is the so-called causal joint?[113] This is an important question not only for ID but for theists of any stripe. Nicholas Saunders began his deep dive into the topic, *Divine Action and Modern Science*, by noting that a coherent concept of special divine action (or SDA) is needed for any theology but that the problem is that there is no "space" for God in the modern scientific understanding of the world, which sees nature as an "it," not a "thou."[114] If nature is an impersonal machine, seemingly maintaining its existence on its own outside and independent of God, then this problem seems possibly unsolvable. Despite running through several options in his book, Saunders concluded bleakly that only Arthur Peacocke's "whole-part" or "downwardly emergent" conception of action has any plausibility for the theist. His model, which seems to be inherent in his panentheism (the view that the world is in some sense included in God's being, a bridge too far for most ID supporters), "sits on the crossroads between emergence and reductionism."[115] But even Peacocke, Saunders felt, did not provide enough of a solution, and so he concluded by noting that the theist's best option is a broad, regulatory approach to divine providence over nature, in which God is not seen in opposition to nature such that he must act against it at discrete moments in time (of the kind that would be empirically detectable).[116] This would favor theistic evolution far more than it would favor ID, as the latter—as we have noted—hinges on the ability of human scrutinizers to find empirical evidence of divine action in nature. ID would later have difficulty drawing support from theistic evolutionists—especially Catholic Thomists like Francis Beckwith (a former sympathizer with ID) and Edward Feser—precisely because of their overreliance on the mechanical view of nature, which influenced their conception of divine action, and their fidelity to eighteenth- and nineteenth-century design arguments like Paley's.

This all makes the association with Paley ironic for multiple reasons. First, there is his aforementioned modern and mechanical view of nature. Second, as the historian Adam Shapiro notes, Behe and Richard Dawkins may have agreed on very little, but they did share the belief that Darwin and Paley were opponents in the great struggle to identify the source of biological adaptation. This widespread interpretation has been challenged by recent scholarship as a projection of the present onto the past. "It was not until the twentieth century," Adam Shapiro writes, "when a new synthesis of evolutionary thought was held against a caricature of Paley's original ideas, that people began to claim that a scientific argument for design had been refuted."[117] Historians have widely contested this narrative of Paley's decline and Darwin's corresponding ascent in the nineteenth century.[118] For one thing, Paley was already seriously outdated even when he started writing—adhering, as he did, to a relatively younger creation by divine fiat and also to a static understanding of species. Regarding natural theology, Jonathan Topham argues that Paley's position in the Darwin mythos led to an overemphasis not only of his importance but also of the importance of natural theology as a whole (evangelicals, after all, did not have much use for it).[119] Nor was the focus on contrivance the only kind of design argument in vogue in the nineteenth century. Wholesale "creation by law," rather than periodic intervention, was a popular interpretation of design in the world after Laplace's nebular hypothesis became more accepted (a view that could fit with the more providential view of divine action).[120] Both before and after Darwin, design arguments sometimes shifted from particular instances of contrivance and unexplained complexity to a focus on the whole pattern of creation as indicative of design.[121] Even Paley, for his part, seemed to think that natural theology was not that useful in vacuum, and he argued that his books should be read the reverse order in which they had been written, starting with his final *Natural Theology*, running through his *Evidences of Christianity*, a text that sought to prove the truth of Christianity through analysis of the New Testament's miracles, and ending with his first work, the *Principles of Moral and Political Philosophy*.[122] Natural theology could be a starting point, not an end—one travels from it to revealed religion and then to political and philosophical theory. Within *Natural Theology*, Paley was concerned with more than a demonstration of design;

he was also making a theological argument about God's character. His intention, according to Shapiro, "was never simply to show that some designer merely existed. His aim was to do theology: to answer religious questions about the kind of being that could create a world filled with purpose."[123] The nineteenth century showed that design could take many forms, but Paley felt that design by itself was not enough to build a worldview.

The third irony is that ID advocates tried to stop at evidence of design and go no further with theology. They tried to do exactly what Paley did not, and this despite their eager association with Paley and the resulting criticism that they were doing natural theology and not science. While ID advocates were often hospitable to Paley's demonstration of design, many of them denied that ID was a new evolution in natural theology—Dembski in particular. "If intelligent design were a form of natural theology," he wrote, "then intelligent design should be looking at certain features of the natural world and therewith drawing conclusions about some reality that extends beyond the natural world. Is intelligent design doing that? I submit it is not." The key for Dembski was the expansion of knowledge from the natural world into the supernatural. Natural theology may have been useful in its recognition of complexity in nature—and its inference of design from this complexity—but it was undone by its foolish insistence on discerning God's attributes from nature, be it benevolence, wisdom, or omnipotence. "Natural theology," he argued, "was primarily in the business of identifying and expatiating on features of the natural world that provided independent evidence of what revealed or sacred theology already knew about God, namely, that God is powerful, wise, and good." This move was impermissible, for ID should not be concerned with the nature, attributes, or personality of the designer for whom they claimed to have found evidence. Dembski wrote, "The theory of intelligent design . . . is not an atavistic return to the design arguments of William Paley and the Bridgewater Treatises [influential pieces of natural theology in the early nineteenth century]. . . . Paley's approach was closely linked to his prior religious and metaphysical commitments. Ours is not. Paley's designer was nothing short of the triune God of Christianity, a transcendent, personal, moral being with all the perfections commonly attributed to this God." ID advocates with religious commitments—including Dembski—were welcome to identify the designer with the God of their personal faith, but it was "strictly optional as far as the actual science of intelligent design is concerned."[124]

This strict focus on the empiricism of design, walled off and hermetically sealed from any theological or philosophical conclusions that could be drawn from natural evidence, meant that there was a split between the theology of intelligent design and its science. This split had two effects: it protected design from theological counterattacks from skeptics, and it also distinguished design more fully from creationism, which was usually wedded to a Bibliocentric worldview (or, in the case of scientific creationism, one that sought scientific evidence of events described in the Bible, such as the flood).

The Problem of "Bad Design"

ID proponents were loath to discuss God's attributes because it opened them up to strong philosophical criticism from evolutionists. They sought, instead, to soften the insistence on deriving God's attributes from nature, and some writers were reluctant to speculate on the designer's identity at all. Indeed, some ID advocates pulled back even earlier, with winking suggestions that the creative force behind life on Earth could be anything with intelligence, even aliens. Such evasiveness was necessary because there was a fatal danger in scouring nature for evidence of God's character. "For every instance where the natural theologian finds reason to sing God's praises," wrote Dembski, "the natural anti-theologian finds reason to lament nature's cruelty."[125] The problem in discerning God's attributes from nature was that it raised questions about God's character in light of all the violence and horror of the animal world (it was, after all, the violence of nature, such as that of the Ichneumonidae wasp, which lays eggs in its paralyzed but still-living prey, that gave Darwin his doubts).[126] Though ID advocates took great pains to avoid this problem, they were often dragged into the conversation against their will. The skeptic's most powerful counterargument, the argument from "bad design," forced ID advocates to deal with it directly. "Bad design" was not a logical reply to ID but rather a probabilistic one. Evolutionary development often has an ad hoc character to it, for evolution is a "bricoleur" and not an engineer, in the words of the developmental biologist Scott Gilbert. A quintessential expression of the "bad design" argument would be Gilbert's highlighting the Rube Goldberg–like functioning of the human reproductive and waste disposal systems. Gilbert related that his wife, who is a gynecologist, "has to deal with these design flaws on a daily basis." In both his and her words, "an omniscient and benevolent deity would not have made such a poorly designed world."[127]

ID proponents usually took two paths in responding to this argument. The first was that questions of the "goodness" or "badness" of designed things were metaphysical in nature, not scientific, and so had no real impact on design as a scientific theory, reflecting only on the character of the designer. The second was to argue that, appearances notwithstanding, such examples of bad design were actually good design.

Dembski took the first approach, arguing that there was a distinction between optimal design and apparent design. He levied the charge of natural theology back at ID's critics, alleging that they were the ones trying to discern the attributes of the designer from the design. "Not knowing the objectives of the designer," he wrote, critics were "in no position to say whether the designer has come up with a faulty compromise." The question must stay within biology, and "taken strictly as a scientific theory, intelligent design refuses to speculate about the nature of this designing intelligence." The moral component is interesting, but to bring it up is to "[leave] science behind and enter . . . the waters of philosophy and theology."[128] Michael Behe mocked Jerry Coyne for his deployment of the bad design argument, asking with rhetorical acid, "Wow, the great theologian Jerry Coyne has determined that God wouldn't have done it that way—no need for actual evidence that Darwin's mechanism can do the job."[129] Whereas Darwin's faith may have been wrecked by the violence and horror of nature, ID refused to assign a moral quality to nature and thus slyly undercut the greatest objection to theism in history, the problem of evil. Evil is not an empirical concept, so it is meaningless in the context of biology and design.

While ID and "bad design" were certainly logically compatible, ID advocates had a harder time evading the charge when it was presented probabilistically—its more common form. Coyne, for instance, did not simply argue that "God wouldn't have done it that way" but rather that, when taken at face value, the systems found in nature were more accurately predicted and understood by a mechanism like natural selection, which one would expect to build things with an ad hoc and often absurdist, roundabout complexity. A prominent example, in his view, was the "circuitous path of the recurrent laryngeal nerve," which was "not only poor design, but might even be maladaptive." It was a legacy of the desultory history of evolution by natural selection—persisting because it still worked, despite its bizarre structure. The recurrent laryngeal nerve was not alone, either: "Courtesy of evolution, human reproduction is also full of jerry-rigged features."[130] Richard Dawkins asked—when noting that a cheetah seemed superbly designed to kill gazelles, while gazelles were superbly designed to evade cheetahs—"For heaven's sake, whose side is the designer on?"[131] The charge here was not necessarily that a designer would not have created things this way but rather that the hypothesis that, say, the recurrent laryngeal nerve evolved via the ever-present coaxing of natural selection better explained the facts of nature. It is the inference to the best explanation. The ID proponent Michael Egnor agreed with this angle and responded to Coyne's argument in a much different fashion than Behe, going so far as to argue that Coyne was doing "good intelligent design science." Coyne framed the question as an inference to whatever explanation seemed to comport best with the facts, which was how Egnor felt that science should be done—that is, "we look at a biological structure and function, and ask we ask: 'Is this evidence for intelligent design, or is chance or unintelligent natural selection a better explanation?'" That Coyne favored natural selection was no matter, for Egnor argued that it was the method that was conducive to design, rather than the conclusion.[132] The disparity in responses between Egnor and Behe illustrated not only the high degree of variation that existed within ID but also that there was a great difference between Coyne speculating about God's character and Coyne hypothesizing about the likelihood of design versus natural selection when analyzing the evidence of nature.

The strength of the bad design argument was sometimes felt strongly enough that ID advocates replied in kind, stating that—appearances notwithstanding—such examples were actually good design. A typical point of contention was the panda's thumb, which featured in the ID textbook *Of Pandas and People*, where it was included as part of a critique of homology and species classification.[133] Stephen Jay Gould's classic essay on the topic typified the bad design argument. He framed the strange thumb of the panda (which is not a digit but rather an extension of the radial sesamoid bone) as not so much an argument against design as it was evidence for evolution and its haphazard nature.[134] Douglas Axe resisted this view and argued—similar to Behe and Dembski—that the question of the good or bad design of the panda's thumb was beyond the scope of science, stating, "we're free to form opinions on these matters, but they're nothing more than that." Axe did, however, venture a bit further into the narthex of natural theology when he admitted, "My opinion, for those interested, is that the giant panda is yet another example of something perfect—something that is exactly as it should be."[135]

Sometimes the defense of nature's perfection could be pushed to the extreme, as when the ID chemist Marcos Eberlin argued that the function of the appendix in "repopulating the GI tract after diarrhea" was evidence of design and foresight because the appendix's "location is perfect from a hydraulic engineer point of view."[136] Jerry Coyne responded by arguing that it was a misconception that "to be vestigial, an organ cannot have a function." "But what is certain," he continued, "is that IDers like Eberlin . . . are suffering from an extreme failure of the imagination in saying that diarrhea evinces an Intelligence On High."[137] The larger point is that in speculating about the perfection of biological systems like the human appendix or the panda's thumb, ID came the closest to its roots in natural theology. It was here that the specter of William Paley loomed most clearly—in the beautiful adaptation of the creature to its environment, of its perfect and undeniably beneficent design, and of the character of the mysterious designer that is transcendent of the universe. Even so, ID advocates largely resisted the temptation to do design-based theology, which had two important consequences for their relationship to creationism. On the one hand, ID's lack of doctrinal interest angered and alarmed creationists who felt they overlooked the Bible, but on the other hand, design's doctrinal neutrality meant the "big tent" could welcome creationists into its movement with relative ease, provided they checked their dogmas at the door.

Design and Doctrine: What Did Young-Earth Creationists Think About Intelligent Design?

Though ID advocates disclaimed any religious motivations in their approach to science, they did not hide their religious convictions (which were mostly, though not exclusively, Christian). Despite the public denunciation of natural theology and the insistence that the designer of ID theory was not necessarily the Christian God, or even a god at all, most of the major ID figures were Christians themselves, who saw in ID a way to unseat atheistic naturalism as the dominant philosophical paradigm in science (though there were more ID sympathizers drawn from the Catholic fold than was usual for YEC). The very subtitle of Dembski's book *Intelligent Design* called ID the "bridge between science and theology," and within, he argued that ID was the "completion" of science and suggested that Christians "follow [Karl] Barth's example and use Christology as the lens not just for understanding the whole of Christian theology . . . but even more ambitiously for understanding all the various disciplines," including science.[138] Critics were quick to recognize this, too. Forrest and Gross quoted Dembski's claim that ID is a form of the "Logos theology of John's Gospel," and then contended, "statements such as these . . . betray the disingenuousness of Dembski's [past] statement that 'design has no prior commitment to supernaturalism.'"[139] Phillip Johnson's religious commitment and language were so clear that the paleontologist Niles Eldredge even praised him for not concealing his convictions, as Eldredge alleged that scientific creationists had done.[140]

The embrace of religion, and the simultaneous assertion that design was science unencumbered, convinced most critics that ID theorists were talking out of both sides of their mouths, trying to have their religious cake and their science, too. Scrutiny of their statements about natural theology, science, and revelation, however, seems to indicate that ID proponents were not necessarily duplicitous in this regard but rather functioning with a split mind—one side embracing revelation *tout court* and the other embracing nature and empiricism. The inconsistency came from the separation of these two modes of knowledge and the almost total lack of interest in harmonizing them. What of the major Christian doctrinal tenets of creation, fall, and redemption? How did these beliefs fit into intelligent design? The absence of ID-themed answers to this question resulted in frequent criticisms leveled at IDers by creationists who found ID inadequate. For the creationists—such as Henry Morris, Ken Ham, and Norman Geisler—ID's problem was not only insufficient religion but also that it failed to bridge religion and science in a way that made scripture and revelation compatible with natural history.

The philosopher Del Ratzsch, a sometimes critic of design who also presented at the Mere Creation Conference, recounted that Ronald Numbers's history of creationism did "not so much as mention William Paley, *The Bridgewater Treatises*, [or] the natural theology movement." Instead, Numbers focused on the influence of George McCready Price and Seventh-day Adventism, and "in the chapter devoted to Price the whole concept of design is only mentioned once in passing, and the design argument not at all."[141] The reason was that the argument from design, perhaps surprisingly, did not feature prominently in the history of young-Earth creationism, especially not in the world of flood geology.

One of the few creationists to focus on design in that period was Richard M. Ritland (1925–2019), who joined the Adventist-backed Geoscience Research Institute (GRI) in 1960 but whose anti-evolutionary argumentation diverged from the usual creationist focus on geology and turned to emphasize design.[142] "Instead of poking holes in evolutionary theory," wrote Numbers, Ritland "highlighted the purported evidences of design in the natural world." The question of whether such evidently designed features of nature were the result of accident or divine purpose was key, and as he considered what he saw, Ritland privately moved away from the young-Earth creationism espoused by his associates (especially Price), accepted the antiquity of the Earth, and saw the flood as a local event. Animated by increasingly heterodox creationist beliefs, Ritland butted heads with the more traditionalist church president Robert H. Pierson (1911–1989) and eventually resigned from the GRI.[143] His career was something of a microcosm of the greater suspicion that flood geologists sometimes evinced about design arguments and illuminates the skepticism with which creationists of various stripes met ID.

Henry Morris repeatedly cast doubt on the ID enterprise, for he worried that its divorce of reason and revelation would not gain IDers any special favors with the secular scientific establishment and might even backfire because the primary source of knowledge about the natural world should be the Bible. As he wrote, "But the ID people (creation *by* Intelligent Design) insist that these are two different systems and that Intelligent Design is certainly *not* Scientific Creationism—especially not Biblical Creationism. They feel it best to leave the Bible and the Biblical God out of the argument entirely. Some even feel that evolution is okay, provided that it is not atheistic Darwinian evolution." This was perhaps a shrewd tactic, Morris saw, because ID advocates felt that "such flexibility is necessary to get the creation idea into the public arena at all." However, mere exposure was not that helpful to the anti-evolutionary cause, for "if the ID system has to be so diluted as to be acceptable to any religion or philosophy except raw atheism, then why bother?" He expected that ID theorists would receive little welcome, even with their Bibleless critiques. "By ignoring this historical evidence—especially that in the Bible," he wrote, "the Intelligent Design movement alone cannot possibly succeed. In the meantime, it is diverting interest among Christians away from the much more cogent case for scientific Biblical creationism, and thus tragically hindering a true witness for Christ and the Bible."[144] Such criticisms were not new for Morris, either. He had seen the way that an overemphasis on design arguments was caused by a shrinking commitment to the biblical stance against evolution. The curious case of Richard Ritland was not sui generis. When there was an influx of old-Earth ideas into the young-Earth-focused Creation-Deluge Society, Morris wrote, "opposition to the Society's basic premise [flood geology] was first expressed in other complaints." Chief among these other complaints was that "not enough attention was being paid to biological evidences of design." Morris sensed that beneath this complaint, the real issue was "embarrassment at the strong position of the Society . . . on the vital issues of recent creation and the cataclysmic deluge."[145]

Morris was not alone in holding that design often served as a culturally respectable swapping for Bible-based creationism, as other creationists expressed the same fear that ID was merely a stillborn critique of Darwin. Norman Geisler, despite his early embrace of many proto-ID arguments, took umbrage at Michael Behe's contention that it was "silly" to treat the Bible "as some sort of scientific textbook." Geisler riposted, "While the Bible is not a systematic science text on the various sciences, nonetheless, there is no evidence to demonstrate that it is not scientifically accurate when it speaks on matters of origin."[146] Ken Ham—the founder and leader of Answers in Genesis—offered a rather more personal attack on Behe's methodology (and ID in general). "I have also said many times over the years that just teaching about intelligent design is not enough," wrote Ham. "One needs to point people to the true Designer—to the Creator God of the Bible." The proof of ID's inadequacy was in the fruits it had borne, for Ham pointed out that Behe's own son,

Leo, became an atheist "even though he grew up knowing the ID arguments." "This serves as a reminder for parents," continued Ham with his characteristic severity, "not to compromise the Word of God."[147] Scientific critics may have seen too much theology in intelligent design, but many creationist leaders did not see enough.

Conclusion

The metaphysical critique of evolution, indeed of the practice of science itself, exemplified by Phillip Johnson and his attack on naturalism, was the foundational element of ID's philosophy. This led, then, to an ostensibly empirical attack on Darwin, shorn of any biblical or religious language. But ID advocates knew their new paradigm needed a scientific basis too, and this case was made most trenchantly by Michael Behe and William Dembski. However, with a campaign against evolution that was almost all offensive, with only the rudiments of a paradigm offered to substitute for it, critics turned to examine the religious impulses and beliefs of many ID proponents. In doing so, they deduced that Johnson, Behe, and Dembski—as well as many others—were really trying to smuggle God back into the scientific picture à la early modern natural theology. Though ID advocates were hospitable to such figures as Paley, they were reluctant to either admit to doing theology or to devote much time to engaging in it. Rather, for Dembski, "intelligent design is at once more modest and more powerful than natural theology."[148] ID differed from early modern natural theology, as it deliberately moved the question away from determining God's attributes and character from nature, a strategic move to avoid the complications of theological dogma for science. As the biologist and former Catholic priest Francisco Ayala argued, it was "well-meaning, if naïve, arrogance" on Paley's part to assume that he could identify evidence of God's personal attributes in nature.[149] In restricting the investigation only to nature, ID circumscribed much of the project of natural theology and therefore inoculated itself against some of the most pressing problems that have beset natural theological thinking—chiefly the problem of evil and the "bad design" argument.

This strategy, however, had its consequences, for it illustrated a sort of bifurcated thinking in design: nature and revelation were wholly distinct and hermetically sealed off from each other. Nature, in fact, might even be a kind of machine that exists independent of God's constant sustaining power. Nature and revelation might inform each other (as Christianity certainly teaches the doctrine of creation and the Fall of Man), but they could not directly communicate. The split between nature and faith therefore earned the enmity of many young-Earth creationists and showed where ID and YEC differed most greatly. It also exasperated critics of design, who often sensed that ID advocates were either inconsistent or deceitful in their insistence that design was only scientific and not religious. Conversely, it strengthened ID considerably as a political movement, for the lack of doctrinal commitment meant that any and all crusaders against Darwin could find a home in the big tent, regardless of their beliefs about the flood or the Earth's age. This flexibility meant that it was as politics that ID would find its widest success—and not only as a scientific challenge to Darwin but also as a populist reaction against the powers that be, one largely rooted in the internet and new technologies of communication and oriented against the scientific establishment that anti-evolutionists felt had lost touch with the people and their values.

3

Politics

Conservatism and Darwinism

To revise a popular movie saying, you either die a liberal or live long enough to see yourself become a conservative. William Paley chose the first option. Though he was not an especially conservative thinker in life, after his death, he became widely reviled by the English lower and middle classes as a symbol of the stuffy elites whose roseate worldview supported a static and hierarchical universe. This was a somewhat unjust drop in reputation, for as D. L. LeMahieu notes, Paley evinced an anti-aristocratic tone in many of his political writings, and he "contributed to the growing polemic against aristocracies that arose in both Europe and America in the late eighteenth century." However, Paley never wanted to overthrow the established order, preferring the belief that even if "private property embodied unsavory elements, there must be some reason, some larger purpose or design, which would justify its continuing existence."[1] That the hierarchical social order of nineteenth-century Britain might be divinely ordained—indeed designed—made Paley an apparent symbol of the restrained conservatism that characterized Britain in the era after the French Revolution, especially since he was so prominent at Cambridge University, then the bastion, along with Oxford, of the aristocracy (and, of course, Darwin's alma mater during his gentlemanly youth). Radicals like the surgeon and social reformer Thomas Wakley (1795–1862), writes the historian Adrian Desmond, sought to use new developments in morphology to "confront the Oxford-educated medical elite, with its static creationism and Paleyite natural theology." Indeed, "working-class atheists . . . lampooned this 'design' argument simply because it supported that poisonous 'monster' priestcraft, and with it the iniquities of the undemocratic state."[2] Likewise, the combative English pamphleteer William Cobbett (1763–1835) excoriated Paley for his political elitism, writing, "I call upon Blackstone and Paley to come forth from the grave, vindicate their writings, and tell, if they can, of what use is a House of Commons, except that of amusing the unthinking mass of the people with the idea that they are represented."[3]

Design, the design argument, and intelligent design would never really shake off this association with conservatism—both in England and later in the US. ID continued this political tradition of design, for conservatism formed one of the central pillars of the wider intelligent design movement. ID's social worldview, in fact, was one in which the "culture wars" were of critical importance. Phillip Johnson's *Reason in the Balance*, which his protégé John Mark Reynolds considered Johnson's most important work, focused not on the science of evolution but on its social and philosophical basis.[4] The issue, as we have seen, was one of metaphysics. The "grand metaphysical story" of modern science held Darwinism at its center, and this world-picture was fundamentally incompatible, in Johnson's mind, with theism and a conservative worldview.[5] To save the culture from the corrosive impact of Darwinism on civic morality and ethics, Darwinism—being the foundation and justification of secular worldviews—would have to be unseated and replaced with something more amenable to a conservative moral outlook. For Johnson, this alternative position was "theistic realism."[6] The main think tank behind intelligent design—the Discovery Institute—christened its ID wing as the "Center for the Renewal of Science and Culture," highlighting the hope for moral resurrection (the center later dropped the word "Renewal"). The center was given life after a 1995 conference on the "death of materialism and the renewal of culture," which helped, in Discovery President Bruce Chapman's words, to "mobilize support to attack the scientific argument for the 20th century's ideology of materialism and the host of social 'isms' that attend it."[7]

The conservatism of intelligent design, however, was of a different kind than that which was normally associated with young-Earth creationism. While creationism mostly mixed fundamentalism, populism, and the grassroots morality of the US South and West, ID's conservative ancestry was far more urbane, effete, and elitist (in many ways, it was a social throwback to the old-Earth creationism of nineteenth-century theologians like Charles Hodge, 1797–1878). The origins of ID-conservatism are found not in the countryside but in the country club. Cerebral, academic, and often Catholic, ID-conservatism's progenitors were not fire-and-brimstone circuit riders with tent revivals on the mind; rather, they were intellectuals like Richard Weaver, Russell Kirk, and William F. Buckley. Whereas creationism was usually evangelical and Protestant, ID (though it housed many evangelicals) featured with them mainline Protestant figures, as well as Catholics, Eastern Orthodox, and Jewish agnostics and theists. To the traditionalist conservatives, anti-Darwinism took on a far broader and more general tone—indeed, one might say almost ecumenical, expanding across denominations, and based primarily in social criticism and philosophy rather than in biblical interpretation. The path of this conservative stream, which would find its greatest scientific expression in ID, can be followed from the traditionalists of the immediate post–World War II era through the rise of the so-called neoconservatives in the 1970s–early 2000s, and finally through the "theocons" (in Damon Linker's phrase) of the 1990s and early 2000s.[8] The political coalition constructed to cover all these different groups was necessarily one with, as we have seen, minimal doctrinal commitments. Thus, the conservatism that lurked beneath intelligent design could contain multitudes: a broad alliance of traditionalists, nonreligious conservatives, irascible agnostic skeptics, and even a fair number of young-Earth creationists. If they could not unite around the age of the Earth, they could and would rally around common conservative social causes. This they did, but eventually ID's occasional political victories, notably its penetration into public school science curricula, would prove to be pyrrhic.

Creation in the Country Club: Traditionalists and Evolution

In the early 1840s, the publisher Robert Chambers (1802–1871) was recuperating from depression and a "diseased" mind in St. Andrews, Scotland. In between games of golf, Chambers wrote a book called *Vestiges of the Natural History of Creation*, and its 1844 publication was a watershed in the history of evolutionary theory. Before Chambers, James A. Secord argues in *Victorian Sensation* (his book about *Vestiges*), evolutionary ideas were restricted to radicals, anarchists, and malcontents. Socialists, communists, and materialists of all stripes may have adhered to it, and it may have reached some common currency on the streets of England's rising industrial towns of Manchester and

Birmingham; but it had not been welcomed into polite society. *Vestiges* would lay out a cosmic vision of evolution—of the development of plants, animals, humanity, the planets, the cosmos, all the universe—and because Chambers feared reprisal, he remained anonymous when the book was published.[9] He feared rightly; *Vestiges* ignited a firestorm. Adam Sedgwick (1785–1873), the great geologist and Darwin's mentor, supplied the literary equivalent of a defenestration in the *Edinburgh Review* (and theorized in a letter that the book's numerous problems indicated "the work is from a woman's pen"—he suspected Lord Byron's daughter Ada Lovelace). William Whewell (1794–1866), the legendary polymath who coined the word "scientist," was dismissive of Chambers's speculations.[10] Richard Owen (1804–1892), the esteemed but prickly paleontologist who invented the word "dinosaur" and had the temperament of one, too, refused to give it the time of day. Owen was ambiguous in public response to Chambers, but privately he wrote to Whewell that refuting the *Vestiges* would give it "an importance calculated to add greatly to its mischief."[11]

Chief among the shocks of *Vestiges* were the social and moral consequences of its theories. Sedgwick worried that it reduced humanity to "the children of apes and the breeders of monsters" and that Chambers "annulled all distinction between physical and moral." The result could only be "rank, unbending, and degrading materialism."[12] Chambers was not a materialist, however, and he hoped for philosophy to take up the baton where science stopped, arguing that there must be "a First Cause to which all others are secondary and ministrative, a primitive almighty will, of which these laws are merely the mandates." Chambers was at home in a theistic cosmos, and in fact, when discussing adaptations, Chambers simply pointed the reader to Paley's *Natural Theology*.[13] Altogether, *Vestiges* did succeed in moving discussion of evolution out of the radical fringes of England and into aristocratic society, so that a gentleman naturalist like Darwin would be able to wade into the debate over the origin of species in 1859 with the climate of opinion a good deal more favorable.[14] Chambers died in 1871 and was buried inside the Church of St. Regulus, in the ruins of the old St. Andrews Cathedral.

Just over a century after the publication of *Vestiges*, an American doctoral student would walk the same grounds as Robert Chambers—and perhaps even pass by the latter's grave in the ruins of the cathedral—as he contemplated the history of conservative thought in the Western world. Russell Kirk (1918–1994), who would become the first American to be awarded a doctorate of letters from the University of St. Andrews, eventually compiled one of the twentieth century's greatest tomes on conservative thought, *The Conservative Mind*. Though he did not write much on Darwin, Kirk clearly found Darwin's theories irreconcilable with conservative principles. Darwinism was chief among "scientific doctrines . . . which have done so much to undermine the first principles of a conservative order."[15] Conservatism, for Kirk, could be understood as a multilayered set of principles. There were five in total, but two are most important here. First, as summarized by the conservative philosopher Roger Scruton (1944–2020), "a belief in a transcendent order, which Kirk described variously as based in tradition, divine revelation or natural law," and, second, "a conviction that society requires orders and classes that emphasize 'natural' distinctions."[16] It is easy to see how evolution—especially in its Darwinian forms, which usually (at least in the minds of its critics) invited nontheistic materialism—militated against both principles. The naturalistic metaphysic behind Darwinian evolution eroded the foundation of transcendent order, and therefore of divine revelation and natural law, and blurred the taxonomic boundaries between species, which made it more difficult to isolate and understand the "natural distinctions" that were central to Kirk's worldview. The perceived radicalism at the heart of evolution, the legacy of Chambers's and Darwin's ideas, compelled Kirk to argue that no materialistic version of evolution could be compatible with a worldview that valued tradition. "There is no place in the Darwinian worldview," the philosopher and ID proponent Robert Koons observed, "for those 'permanent things' that Russell Kirk postulated as the foundation of conservative thought."[17] Kirk hoped one day to write a sequel to *The Conservative Mind*—titled either *The Age of Humanism* or *The Humane Tradition*—and planned to mimic the style and structure of Plutarch's *Lives* (with each "life" paralleled with an antithetical one). He would have made Charles Darwin one of his principal figures, giving him a full chapter, but Kirk never completed the work.[18]

Where Kirk left an opening, however, other traditionalist conservatives rushed in. Richard Weaver (1910–1963), a Catholic influenced by the Southern Agrarians, was likewise skeptical of evolution and

Darwin. "Ideas have consequences," asserted Weaver in a book of the same name, whose title and thesis would become one of the mainstays of the conservative movement in the US. As the historian George Nash writes in his still-definitive work on the history of US conservatism, Kirk and Weaver (and many others) believed that one could trace the origin of modernity's ailments through the history of its ideas. For these conservatives, the ills of modernity were at root "an intellectual error."[19] Darwinism, in Weaver's view, was one of the deadliest of modernity's pathologies, for it demoted humanity to the animal realm.[20] Kirk praised *Visions of Order*, where Weaver turned his guns most forcefully on Darwin, for its assault on "the 'presentism,' scientism, and democratism that are subverting the high old order of our civilization and our human dignity."[21] Toward the end of the book, Weaver admitted the daunting task he had set for himself, writing, "I recognize that any layman's criticism of the theory of evolution will appear to most people today as reckless." Because the testimony of the sciences was well-nigh universal in favor of evolution, Weaver proceeded with caution and directed his attention toward the metaphysical apparatus behind evolution, rather than the facts of the theory itself. For Weaver, "The theory of evolution can be viewed as a form of the question-begging fallacy. It demands an initial acceptance of the doctrine of naturalism before any explanation is offered." Press a scientist to answer how the diversity of life came about, and their answer must invariably be "the proximate method which nature would use, *assuming that nature is the only creative force that exists.*"[22] As we have seen, Phillip Johnson would make this argument about a priori metaphysics his primary riposte against Darwin nearly fifty years later. Weaver also expressed doubts about whether tiny modifications over eons of time sufficiently explained the complexity and function of an organ like the eye, and he concluded that it is "the notoriety, if not the reasoning, of the evolutionists [that] has caused it to be very widely accepted that man is one of the animals."[23] Weaver's ideas found home much later in the ID world. The ID advocate Robert Koons turned to Weaver and followed his contention that the layperson was competent to judge the veracity of evolutionary claims, provided that one recognizes its philosophical backing.[24] David Klinghoffer cited Weaver for ammunition against both atheistic and theistic evolution. According to Weaver, the only way to eliminate the "conflict" between science and religion, regarding evolution, was when one side "gives up."[25]

Kirk and Weaver were hardly alone; several other conservative philosophers took up the cause. Eric Voegelin (1901–1985) traced the philosophical view of evolution past Darwin to Immanuel Kant and contended that while Darwin was an expert in empirical biology, he was woefully deficient when it came to the philosophical questions and implications of his theory. For Voegelin, the evolutionary movement's "distinct anti-Christian, secularist flavor" was the "dynamic factor" in the triumph of Darwin's thought. The theory of natural selection, then, had "a popular success and became a mass creed for the semieducated."[26]

The sociologist Robert Nisbet (1913–1996) lauded what he saw as the "revolt of younger scientists" against the "Darwinian mandarinate" and stated that new discoveries in the varied fields of molecular biology, geology, and astronomy (among others) testified to the unwelcome fact that "between the essential principles of evolution which Darwin gave to the world and the principles of evolution which recent research in several areas has yielded, there is nearly total conflict." However, Nisbet, like later ID proponents, professed no allegiance to creationism and declared "what can be leveled against the mandarinate is the myth . . . that Darwinism and evolutionism are identical and that any criticism of Darwin is perforce a criticism of evolution and must come from a creationist."[27]

Lastly, the grand strategist of postwar conservatism, William F. Buckley (1925–2008), also professed numerous doubts about the validity of evolution. Though he mentioned Darwin only in passing in his early classic *God and Man at Yale* (focusing instead on the supposed incompatibility of religion and science in the field of sociology), Buckley would target evolution repeatedly throughout his long life.[28] Outliving many of the other traditionalist conservatives of the midcentury, Buckley went on to support intelligent design directly, holding a 1997 debate on his PBS show *Firing Line*, where he, Phillip Johnson, Michael Behe, and David Berlinski squared off against Eugenie Scott, Michael Ruse, and Kenneth Miller.[29] Only one year before he died, Buckley wrote a column in *National Review*—the magazine he founded in 1955 to help fuse together the disparate traditionalist and libertarian wings of US conservatism—in which he extolled the virtues of Phillip Johnson's work and wrote that "the intelligent liberal community should not impose on anyone a requirement of believing that there is only the single, materialist word on the subject [of evolution]."[30]

Neocons to Theocons: Late Twentieth-Century Conservative Critiques of Darwin

The traditionalist conservatives would find curious congruence with the so-called neoconservatives, several of whom also embarked on anti-evolutionary excursions. Though the target was the same, the impulse was different, as neoconservatives tended to be secular and Jewish rather than religious and Catholic. The meaning and content of neoconservatism is the political equivalent of the gordian knot, and so Justin Vaïsse, in his sweeping history of the topic, opted not to define the word but to illustrate three "phases" of the movement's existence. In the first era, starting in the 1960s, neoconservatives were disillusioned Trotskyists, typically Jewish intellectuals from New York who had gone to the City University of New York, such as Irving Kristol, Norman Podhoretz, and Nathan Glazer (1923–2019). The neoconservatives crystallized around a series of periodicals—Kristol's *The Public Interest* and Podhoretz's *Commentary*, among others—intended to counter the influence of the supposedly irredeemable *New York Review of Books*. The second and third phases of neoconservatism, for Vaïsse, revolved more around foreign policy—most notably in the Reagan years and the close of the Cold War, as well as in the turn of the twenty-first century with the Bush Doctrine and the global "war on terror."[31] The neoconservative interest in democracy, liberal human rights, and its relative lack of explicit religiosity set it far apart from the traditionalist conservatism of figures like Kirk or Weaver, both of whom harbored deep suspicion of democracy. And yet, anti-evolution was prevalent in neoconservatism as well.

The early opus of neocon anti-evolution came in the form of Gertrude Himmelfarb's *Darwin and the Darwinian Revolution*, which was published on the Darwinian centennial of 1959. Himmelfarb (1922–2019), Irving Kristol's wife and an authority on the history of Victorian Britain, evinced little overt evidence of neoconservative ideology in *Darwin and the Darwinian Revolution* but did feature several criticisms of evolution toward the end of the book. (Himmelfarb conflated natural selection and evolution, for she argued that Darwin did not just accept evolution at face value and propose natural selection as the mechanism but that he tried, and needed, to prove both.) She also targeted the same perceived weak points that anti-evolutionists had always done: the lacunae in the fossil record, survival "competition" serving as an inadequate metaphor for nature, and the implausibility of the unguided achievement of complex life, "delicate machinery," at the macro and micro levels. Despite the universal acceptance of evolution and its Darwinian framework of natural selection and descent with modification, Himmelfarb concluded that "in each generation a small number of reputable scientists revived the 'antiquarian' [anti-evolutionary] controversy, reminding their colleagues of Huxley's warning about truths that begin as heresies and end as superstitions." She highlighted the same accusation of metaphysical bias that has been made by most anti-Darwinists, alleging that the scientists committed to Darwin did so in "more an act of faith than of demonstration" and that—in the words of the passionate neo-Darwinian August Weismann (1834–1914)—"it is inconceivable that there could be yet another [principle] capable of explaining the adaptation of organisms, *without assuming the help of a principle of design*." As in ID, the worldview was the thing. Metaphysics determined physics (and biology), not vice versa.[32]

Irving Kristol (1920–2009) expressed Darwinian doubts as well. In a piece in *The New York Times*, Kristol lamented the endless war of "science versus religion" that had been abetted by creationist court conflicts in the 1980s and called for more circumspection when it came to teaching evolution. "Though this theory is usually taught as an established scientific truth," he charged, "it is nothing of the sort." He viewed the in-house disagreements among evolutionary biologists as evidence of a state of conflict within the field and argued that it was disingenuous to overlook such controversies and then foist evolution on a religious populace, particularly since "the current teaching of evolution in our public schools does indeed have an ideological bias against religious belief—teaching as 'fact' what is only hypothesis." Kristol had no patience for creationism, though, and felt that setting creationism alongside evolution in school would be a "booby-trap." Instead, he wanted a modest reappraisal, so that Christians in schools would not feel they were under attack by a cadre of scheming evolutionists dead set on overthrowing traditional values. "If evolution were taught more cautiously," he pleaded, "it would be far less controversial."[33]

Kristol's piece elicited a confused response, for it was clear that he was no creationist (or even particularly religious)—so why attack evolution? The biologist Douglas Futuyma, in a letter to the editor, wrote that Kristol's column "could well have been written by a creationist, and bears no resemblance to the current understanding of evolution by biologists." He reiterated the point that evolution was compatible with any and all religious belief systems, provided those systems are not based on "an interpretation of the Bible as a literal statement of history."[34] Futuyma notwithstanding, none of these figures based their anti-evolution on a literal interpretation of the Bible. The difference in the traditionalist conservative, and later neoconservative, attacks on evolution was a widespread belief within those movements that evolution truly could not coexist with religion, regardless of whether that religion was based on biblical texts. The conflict was deeper—one of worldviews, metaphysics, and ethics.

Arguably the most well-known attack on evolution in a neoconservative journal was David Berlinski's rhetorically dazzling essay "The Deniable Darwin," which was featured in Podhoretz's *Commentary* in 1996. Berlinski, with a novelist's flair for similes and metaphors, castigated what he viewed as complacency and a lack of imagination in contemporary evolutionary biology. He condemned evolution as "a materialistic theory" without need for any deity.[35] Criticizing evolution from the usual angles (complexity, fossil record), Berlinski concluded that evolution was "cherished not for what it contains but for what it lacks. There are in Darwin's schemes no biotic laws, no *Bauplan*, . . . no special creation, no *élan vital*, no divine guidance or transcendental forces."[36] In a follow-up letter, he cleaved to the neocon pedigree in endorsing Himmelfarb's *Darwin and the Darwinian Revolution*. The article provoked a dizzying array of responses, from shock, dismay, and outright ad hominem from Richard Dawkins and Daniel Dennett to nuanced critiques from H. Allen Orr and Arthur Shapiro and to adulation from Phillip Johnson and Michael Behe.[37] When Dawkins and Dennett lambasted Berlinski for being a creationist, he fought fire with fire and questioned their intelligence—or, in his words, Dawkins's "alarming logical deficiency" and Dennett's "curious impression that the best rejoinder to criticism is a robust display of personal vulgarity."[38] Berlinski roundly denied any personal religious inclination, writing elsewhere, "Here it is, an inconvenient fact: I am a secular Jew. My religious education did not take."[39] To further muddy the waters, he later published an equally incendiary attack on intelligent design (this despite being a fellow at the ID think tank, Discovery Institute).[40]

What motivated the neoconservative hostility to Darwin, if it was not based on either biblical literalism or Catholic metaphysics? When Futuyma stated in his reply to Kristol that only literalist fundamentalists had a problem with evolution, he overlooked the stated views of his interlocutor and his cohort. So, why did mildly agnostic neocons make a target out of Darwin? The science journalist Ronald Bailey, in the libertarian journal *Reason*, offered a nefarious hypothesis in a 1997 article, just on the heels of ID's surge to prominence. Not long after Berlinski's article, a conference was held by Elliott Abrams's neoconservative Ethics and Public Policy Center in which Kristol, Himmelfarb, and Tom Bethell (a pro-ID journalist) attended lectures by ID leaders Behe and Denton. Bailey wondered if there was a manipulative method to the madness. "The neocon assault on Darwinism," he wrote, "may not be based on either science or spirituality so much as on politics and political philosophy." Instead, he turned to Paul Gross (biologist, ID critic, and "self-described conservative" who cowrote *Creationism's Trojan Horse* with Barbara Forrest), who alleged that the neocon war on Darwin was "a case of tactical politics." Bailey then highlighted Kristol's confessed belief that religion was needed to sustain civic virtue, and he accused the entire neocon movement—including Kristol, Leon Kass, Himmelfarb, David Brooks, and many others—of hoping to discredit Darwin and buoy religious belief in a ploy to maintain a sort of Platonic "noble lie," in which the deluded masses might be kept in line while the elites get on with the business of running the world. He traced this reasoning to the political philosopher Leo Strauss (1899–1973).[41] This was a plausible reading, for—as the historian Edmund Fawcett has noted—religion's utility versus its veracity has been a persistent dilemma for conservatives through the ages. "How can [conservatives] sustain a belief that [they] are convinced society needs," asks Fawcett, "when [they] offer not grounds or evidence for the belief but only a conviction that the common holding of the belief is useful for social order?"[42] Himmelfarb addressed the same point, writing that Darwin's father, Robert, felt that "while he personally was superior to the irrational dogmas of religion, the masses of men required the stabilizing influence of a church." Even figures as skeptical as Thomas Henry Huxley "wrote polemics in favor of Bible reading." Himmelfarb pointed to this as proof that Victorian agnostics were "willing to make concessions to religion in the interests of public morality. They were willing to suspend their own disbelief in order to bolster up other people's morals."[43] This accords well, too, with Kristol's self-application of the adjective "theotropic," despite a reluctance to practice the "decadent Orthodoxy" into which he was born. As he wrote, a prosperous society "needs the energies of the creative imagination as expressed in religion and the arts. . . . Nothing is more dehumanizing, more certain to generate a crisis, than to experience one's life as a meaningless event in a meaningless world."[44]

The religious wing of late twentieth-century conservatism did eventually object to the instrumental view of religion taken by some neoconservatives. The "theocons" espoused many of the same views as their neoconservative brethren, but with a decidedly religious flavor. As Damon Linker tells the story, some conservative thinkers grew frustrated with the apparent neoconservative use of religion as a tactical gambit. These "theoconservatives," as he coined them, coalesced primarily around the popular journal *First Things* and the Catholic convert Richard John Neuhaus (1936–2009).[45] As Linker notes, *First Things* would also play a prominent role in the emergence of intelligent design; however, when the details are scrutinized, it does appear that the conservatives at *First Things* were more ambivalent in their ID support than Linker indicates. That said, they did play a role in getting it off the ground, even if many Catholic conservatives later expressed doubt about its viability.

In *First Things'* inaugural issue, in March 1990, the editors—including Neuhaus—outlined their commitments, writing what the core thrust of the new magazine would be: "for the sake of both religion and public life, religion must be given priority." Public life was not just politics; it was also culture. While the US may seem, and was often labeled, an essentially secular society, "the intention of the enterprise is to advance a religiously grounded public philosophy for this and other experiments in human freedom."[46] It would not take long for this "religiously grounded public philosophy" to include design and its critique of Darwinian materialism. A scant half year after the magazine's inauguration, Phillip

Johnson published an essay in *First Things* titled "Evolution as Dogma," noteworthy not only for its appearance in a major, intellectually respectable conservative journal but also because it antedated *Darwin on Trial* by a year. In the essay, Johnson began by immediately distinguishing his critique of evolution from creationism. "There is a great deal more to the creation-evolution controversy than meets the eye," he began, "or rather than meets the carefully cultivated media stereotype of 'creationists' as Bible-quoting know-nothings who refuse to face up to the scientific evidence." Issues of biblical hermeneutics and Noah's flood were what Johnson termed "side issues." The real conflict was over something larger and deeper: it was about philosophy; it was about worldviews. As Johnson continued, evolution was "based not upon any incontrovertible empirical evidence, but upon a highly controversial philosophical presupposition."[47] That presupposition was naturalism. The issue was not that the evolutionist's starting point was the falsity of the biblical record; it was that their starting point was God's nonexistence. It was here that Johnson identified the real locus of the "culture wars." As he would write a few years later, "When there is radical disagreement in a commonwealth about the creation story, the stage is set for intense conflict, the kind of conflict that is known as a 'culture war.'"[48]

To fight a culture war, one needs a few tools. A magazine is helpful, but a think tank is key. As Fawcett observes, because conservatives felt that the academic world was unfairly hostile to them, an alternate source of intellectual power was needed. The interest in ideas and their consequences, spurred by Weaver, led to "a well-financed Gramscian campaign of intellectual renewal in think tanks, institutes, the media, and universities."[49] Such think tanks proved fertile ground for a separate ecosystem of knowledge and expertise outside its traditional halls in US higher education. The intelligent design movement would find its own in the Discovery Institute, based in Seattle, Washington. Originally founded in 1991 with more local aims (such as modernizing the railway transit system outside Seattle), the institute quickly dedicated itself to the task of cultural combat. One of the cofounders, George Gilder, had long been a voice in the conservative cultural faction, publishing frequent critiques of contemporary attitudes toward sex since the so-called sexual revolution. The author of conservative mainstays like *Wealth and Poverty* (on libertarian economics) and *Men and Marriage* (on the role of the family), Gilder turned toward evolution as well and was frustrated by the Darwinian concepts that were rife in both economics and in ethics (particularly sexual ethics). In *National Review*, Gilder opined, "Almost by definition, Darwinism is a materialist theory that banishes aspirations and ideals from the picture. As an all-purpose tool of reductionism that said that whatever survives is, in some way, normative, Darwinism could inspire almost any modern movement, from the eugenic furies of Nazism to the feminist crusades of Margaret Sanger and Planned Parenthood."[50] Bruce Chapman, who cofounded the Discovery Institute with Gilder, long expressed a similar perspective. It was in a meeting with Gilder, as well as Stephen Meyer, that the trio decided that "Discovery should become the home to the scientific critique of Darwinism, and home as well to intelligent design as an alternative theory."[51] The result was an extension of the institute into a new direction, with the founding of the Center for the Renewal of Science and Culture in 1996. Soon, the center would be the main hub of intelligent design, with Phillip Johnson serving in an advisory, godfatherly mode, while the rising ID stars like Michael Behe, William Dembski, Jonathan Wells, Stephen Meyer, and many others would all serve as senior fellows. The integration of design together with cultural critique would form the basis of ID's conservative conflict with Darwinian evolution.

The Great Oxymoron: Can a Darwinian Be a Conservative?

Are all Darwinians liberals or leftists? Are there no Darwinian conservatives? The intellectual coherence of conservative Darwinism was debated at both *First Things* and elsewhere. While Damon Linker portrayed Richard John Neuhaus as an enthusiastic design proponent, he grew more tepid toward it by the end of his life.[52] In 2007, Neuhaus expressed some frustration with the Discovery Institute fellow John G. West's polemic against Darwinian conservatives (it is noteworthy that this resistance came after the *Dover* case). West charged that the conservatives Charles Krauthammer (1950–2018) and George Will did not take their cultural critiques all the way to Darwin (and indeed, they accepted evolution), but Neuhaus disputed his approach. He cautioned that evolution did not necessarily entail atheism and pointed to convinced Christian evolutionists like Francis Collins as evidence against the conflict thesis. Equating evolution one-to-one with atheism has the effect of conceding "into the hands of those who would perpetuate what they have an interest in portraying as an inevitable war between science and religion."[53] Likewise, when Johnson published his later book *The Wedge of Truth* in 2000, he was met with vigorous resistance by the Jesuit priest Edward T. Oakes (1948–2013), who contended in *First Things* that Johnson's strategy for intelligent design was beset by serious "theological inadequacies. . . . The main problem, at least for a theologian, is that the results are so nugatory."[54] For Linker, resistance like Oakes's was early and trivial, but in the long run ID was not fully accepted by those in the *First Things* orbit. Later, conservative Catholics like Francis Beckwith and Edward Feser would embark on critiques of design from a Thomist perspective. While these rejoinders show that not all evolutionists—or at least those sympathetic to it—were liberals, the incompatibility of Darwinism with conservatism was still central to the ID polemic, just as it was for Richard Weaver and Russell Kirk.

The philosopher Larry Arnhart attempted to unify evolution and conservatism in his books *Darwinian Conservatism* and *Darwinian Natural Right*. In a disputation held in *First Things*, Arnhart tried to persuade ID's leaders of his position and defended three things: the truth of evolutionary theory, the compatibility of evolution with theism, and the compatibility of Darwinian and conservative ideologies. For the latter, Arnhart wrote "Darwinian views of human nature provide scientific support for the traditional idea of natural moral law. Human beings really are naturally social and moral animals, and therefore we can judge social life by how well it conforms to the natural needs and desires of the human animal. Natural law is not a 'myth.' It is a rationally observable and scientifically verifiable fact." Arnhart concluded that teleology in biology, the lynchpin of design and conservatism, was not eradicated by Darwin per se. Rather, while Darwin may have complicated any notion of "cosmic teleology," he actually reinforced the potential for an "immanent teleology" within nature. This immanent teleology could still provide the basis for a natural law assessment of human nature, and even Aristotelian ethics, without recourse to extrinsic design. A moral law thus rooted in nature would not only be compatible but perhaps even necessary for conservatives, and Darwin, in Arnhart's view, could supply such conservative principles. Furthermore, Darwinian evolution indicated that human nature was not infinitely plastic and so was therefore incompatible with progressive political movements that depended on humanity's malleability, such as socialism.[55]

Michael Behe and William Dembski responded to Arnhart with a vociferous negative. Behe began clearly: "I'm sorry to be blunt, but the notion that Darwinism supports conservatism is absurd." For Behe, Darwinism could not be used as the grounding for any moral philosophy, for at most, evolution could only tell what an organism is or how it got to be that way, not how it should be or how it ought to act. Darwinism was too elastic, in Behe's mind, to support any real moral program, conservative or not. "If research shows that humans are selfish, Darwinism can explain that," he wrote. "If science shows we are unselfish, why, it can explain that, too."[56] John G. West both wrote a full-length rebuttal of Arnhart and made him a major target in another work.[57] West focused primarily on Arnhart's contention that order (the pillar of conservative worldviews) was spontaneous in nature and argued that it was impossible that social cooperation or charity could have evolved randomly, for "it results from the intelligent choices of innumerable designers interacting with each other"—that is, human beings themselves. Spontaneous order in nature and human behavior was "misleading." Furthermore, even if Darwin found something like the golden rule in nature, this would not matter because nature did not evolve a moral standard like this "because the golden rule is somehow intrinsically right. It did so because the golden rule ultimately is connected to self-preservation."[58] In the end, as before, the issue was one of metaphysics. The golden rule may be found in nature, but absent a theistic metaphysic, there was no real justification for it other than instrumental usefulness.

The free market, and capitalism more broadly, posed significant problems for ID when it came to Darwin and conservatism, and anti-Darwinian conservatives took different paths in response to it. The seemingly Darwinian nature of the free market—with its competition and its struggle for monetary existence—meant that conservatives had to be careful in explaining how they could reject the one and accept the other. The conservative commentator and atheist George Will framed this dilemma well by highlighting not only the Darwinian connection but also the fundamental aversion to design in libertarian economics, particularly in Friedrich Hayek (1899–1992).[59] Hayek argued that a "naïve mind" would understand "order only as the product of deliberate arrangement."[60] Scorning the planned economy, Hayek felt that the free market was instead the result of "spontaneous order."[61] The similarity between the language of economic design and biological design—and the insufficiency of both—led George Will to reject them as two sides of the same failed coin. Will quoted Matt Ridley, who asked, "If life needs no intelligent designer, then why should the market need a central planner? Where Darwin defenestrated God, [Adam] Smith just as surely defenestrated Leviathan." Later in the same work, Will called ID "intellectual fudge."[62]

Because the free market was a fundamental part of twentieth-century US conservatism (made part of the broader movement through the "fusionism" promoted by Frank Meyer, 1909–1972, and William F. Buckley that helped bridge the divide between Kirkian traditionalist conservatives and libertarians such as Hayek), some ID advocates have attempted to square the circle and rescue capitalism from its liberal and Darwinian trappings. John G. West reversed Will's thesis and resisted the "spontaneous order" view of the free market, arguing that Hayek was suspicious of the adaptability of natural selection to economics, as the Austrian economist contended that social Darwinism had disgraced such ideas. West granted that natural selection was imported into economics in the thought of Milton Friedman (1912–2006) but tried to limit its importance. Regarding Hayek's spontaneous order, West admitted that this was a more "profound link" but argued instead that Hayek's view originated not from Darwin but from David Hume, Adam Smith (1723–1790), and Adam Ferguson (1723–1816). Thus, the spontaneous order of political economy predated Darwin. Furthermore, for West, the "variations" of economic Darwinism were not random as they were in biology, as they were the product not of mutations but of rational choices by *homo economicus*. Rebutting Arnhart by name, West concluded, "Darwinism has offered little genuine support for laissez-faire capitalism." Ronald Reagan's economic views, for West, were less Darwin and more George Gilder.[63]

Along with West, Jonathan Wells disputed the Darwinian nature of the free market, but he had to sidestep the paradox in ID thinking: design in government was bad, but design in nature was good. To save design but defeat planned economies, Wells argued along with Gilder that Darwinian economics would lead to "managed economies" because survival of the fittest was a zero-sum game. Darwinism, for Gilder and Wells, obstructed human creativity because the competition for resources naturally invited an interventionist government to regulate commerce. Conversely, in Gilder's words, creative thought—the lifeline of a market economy—was the result of "love and faith that infuse ideas with life and fire. All creative thought is thus in a sense religious." By stifling creativity in a bid for management, maintained Gilder and Wells, "it is Darwinism, not ID, that encourages the sort of government interference that conservatives abhor."[64]

Not all anti-evolutionary leaders have shared this conclusion, though resistance to capitalism as a Darwinian form of economics was more frequently found in the fundamentalist, creationist wing than it was in ID. William Jennings Bryan, as we have seen, equated Darwinism with industrial capitalism and made the ravaging of the working class by robber barons a key platform in his populist campaign. He was, however, not the only creationist to view capitalism as irredeemably Darwinian. The creationist horticulturist Walter Lammerts (1904–1996), one of the early leaders of biblical creationism and flood geology with Henry Morris, was likewise suspicious of laissez-faire capitalism. As Ronald Numbers documented, Lammerts "broke with the conservative mold of the stereotypical creationist. . . . He strongly supported civil rights, conservation, and progressive social legislation and abhorred John Birchers and other extreme right-wingers." He praised progressives like Robert LaFollette who "gradually modified capitalism" and improved it from the harshness of the Gilded Age, and he felt the polemical attempt to tie Darwin to communism, such as that made by George McCready Price, was unsupportable.[65] Not to be outdone, Henry Morris condemned the Republican Party for its devotion to the free market at the expense of Christian principles such as care for the poor and charity. In his late work *The Long War Against God*, Morris wrote, "It is a mistake to assume, as many do, that political 'conservatism' is necessarily compatible with biblical Christianity," and he argued that Robert Welch (1899–1985), founder of the John Birch Society, was a "strong evolutionist." Moreover, "many leaders in the present-day Republican Party . . . are really the spiritual heirs of the nineteenth century social Darwinists. Most of them are firmly committed to evolutionism and the amassing of great fortunes by whatever methods will succeed in the economic struggle for existence. The recent betrayal of the 'religious right' is a painful reminder of this fact to disillusioned Christians." Morris found Reagan disappointing and believed that the Republican Party simply used and discarded conservative Christians to sweep into power and "restore a greater degree of Darwinist laissez-faire capitalism." In reality, "most political 'conservatives' today are still evolutionists." Morris agreed with many of the aforementioned thinkers that a conservative could not be a Darwinist, but instead of trying to rescue the free market from its liberal trappings, he became more suspicious of capitalism.[66]

Whether a conservative could be a Darwinian or not, there was nevertheless a clear trend of anti-evolution in every major stream of US intellectual conservatism (with the exception of libertarianism, though, along with Kirk—no apologist for the free market—one might argue that libertarians were not conservatives; after all, Hayek once wrote a postscript called "Why I Am Not a Conservative").[67] Whether conservatives were traditionalists, neoconservatives, or theoconservatives, anti-evolution was a thread that ran through each movement. However, for these critics, less important was Jerusalem than was Athens—biblical literalism had never been the basis of either the conservatism of Catholic figures such as Weaver and Kirk or that of secular Jewish intellectuals such as Kristol or Berlinski. In fact, when biblical literalism was prioritized, it sometimes caused a repudiation of the free market emphasis, as Morris showed (though this was not a frequent phenomenon and was restricted more to some of the creationist leadership and not its popular supporters). Instead, the paramount issue was one of philosophy and morality and the apparent (at least to these conservative figures) incompatibility of the blurred boundaries and slippery morality of evolution, as well as the erosion of biological and social distinctions on which conservative order depended, especially on family. A movement based on metaphysics could be a broad basis for a political alliance. However, a truly effective national movement could not be composed only of intellectuals, academics, and elites. The anti-evolutionists covered here have been largely confined to the ivory tower or were lone voices in the wilderness crying out against the consensus. What about the populist wing of conservatism? How did the big tent of ID relate to those who were often out of step with its ideological minimalism?

Elites and Populists: ID's Difficult Balance

In the early 1990s, the scholar Christopher Toumey produced a rare ethnographic study of grassroots, cultural creationism, wherein he documented the social and political beliefs about evolution in a community of creationists in North Carolina's Research Triangle. For Toumey, creationism's social and political worldview matters just as much as its science, and he describes it holistically as "a system of cultural meanings about both immorality and science that helps fundamentalist Christians make sense of the realities, anxieties, changes, and uncertainties of life in the United States in the late twentieth century." The moral decline of the United States was extremely important to these groups, and the culprit for the degradation of contemporary life could be identified in a nefarious secular foe. Writes Toumey, "The core of the New Religious Right's moral theory is the fear that a conspiracy named Secular Humanism causes the evil and immorality that permeates the modern world. Creationism is an exercise in applying that theory to the question of origins, by charging that the idea of evolution is intimately connected to Secular Humanism." While the populist creationism he documented oriented itself around science, it was driven by a social and cultural disposition that expressed itself in a form of political anxiety. "Creationism, then, is much more than a narrow doctrine extrapolated from a handful of biblical verses," writes Toumey. "It represents a broad cultural discontent, featuring fear of anarchy, revulsion for abortion, disdain for promiscuity, and endless other issues, with evolution integrated into those fears."[68] This form of creationism, and the New Religious Right that incubated it, was far less elitist and aristocratic than the traditionalist conservatism of Kirk and company. As George Nash describes it, "whereas the traditionalists of the 1940s and 1950s had largely been academics in revolt *against* secularized, mass society, the New Right was a revolt *by* the 'masses' against the secular virus and its aggressive carriers in the nation's elites."[69]

This theme would be taken up in John H. Evans's provocative book *Morals Not Knowledge*, in which he argued that most of the "conflict" between religion and science in the United States was not driven by epistemological difference but by moral disagreement. Following Peter Harrison, Evans argues that creationists of all kinds were not hostile to science as science as much as they were hostile to science as a perceived liberal Trojan horse, using evolutionary theory to undo the morality of the US rural and working class. "While such varied positions as Young Earth Creationism, Old Earth Creationism, Day-Age Creationism, Gap Creationism, Progressive Creationism, and Intelligent Design can more or less be treated as 'creationist,'" he writes, "it is their qualities as moral criticisms of Darwinism that most tightly bind them together, both historically and sociologically speaking. The concern of creationists in each of their historic incarnations is that when you teach evolution, you are implicitly teaching a certain philosophy at the same time, and that this philosophy undermines some forms of morality."[70] Even going back to *Scopes*, populist resistance to Darwin has been wrapped up in the decline of rural life in the US, which has coincided with increased presence of Darwinism in schools—a correlation that is seen by creationists as a causation.[71] Evans does note, however, that intelligent design was a bit of a different animal in this regard, being one of the few instances in which the epistemological "knowledge-conflict" was often prioritized over the moral one. As he writes, "this is one place where both sides are talking about knowledge."[72] That said, ID still made moral critiques of Darwinism and scientists in general, most notably in the later work of Phillip Johnson (in books like *Reason in the Balance* and *The Wedge of Truth*). So, while ID's struggle against Darwin was primarily an intellectual one, it was not exclusively so, and ID advocates' moral crusade would be a primary point of agreement with populist conservatives who might otherwise have had little interest in its cerebral arguments about metaphysics and scientific paradigms.

No one movement typified this populist attitude more than paleoconservatism. Nash describes the paleoconservatives as a nationalist, "America First" movement more closely affiliated with the Old Right, the isolationist conservatives who predated the Buckleyite consensus. Though their forebears were the Old Right, the "new" to which paleocons were hostile was not the New Religious Right but rather the neocons, whom they viewed as liberal "interlopers."[73] The brain behind the paleo antipathy to Reagan, Bush, and the neoconservatives was Pat Buchanan. As chronicled by Fawcett, it was Buchanan who coined the term "silent majority" for Nixon, and it was Buchanan who made the mainstream media a liberal boogeyman, an idea popular with grassroots conservatives because "people know it's true." "Few," writes Fawcett, "were as skilled at framing arguments in simple terms that all but his liberal opponents, as it seemed, could understand." In fact, Buchanan's paleoconservatism might be the dominant thrust of political conservatism in the post-Bush era. "If *Paleo* were a movie," Fawcett suggests, "Trump would star and Buchanan would get writer's credit."[74]

While conservatives of different persuasions often hated each other, anti-evolution was one of the few common links between each group. Paleoconservatives, and especially Buchanan, were no different. "Darwinism is headed for the compost pile of discarded ideas because it cannot back up its claims," wrote Buchanan in 2005. "It must be taken on faith. It contains dogmas men may believe, but cannot stand the burden of proof, the acid of attack or the demands of science."[75] For the far-right, populist website *WorldNetDaily*, Buchanan wrote a no-holds-barred attack on Darwin in 2009, linking him to every evil thing in human history—from racism to Nazism to communism—and then extolled the virtues of a book by Eugene G. Windchy, who argued that Darwin stole his ideas from Alfred Russel Wallace. Buchanan concluded with, "Darwinism is not science. It is faith. Always was."[76] In a 1996 exposé of Buchanan's anti-evolutionary views, *Time* magazine's Michael Lemonick noted Buchanan's popular appeal and the influence he held over public school boards. "He's not descended from monkeys," wrote Lemonick, "and he doesn't think children should be taught that they are." Lemonick worried that creationists, in the wake of 1987 and *Edwards v. Aguillard*, were becoming slyer in their quest to gain classroom hearing. Despite the constant stream of rebuttals from scientists, "conservative local school boards still use [old creationist arguments] to justify equal-time requirements for ideas like the 'intelligent-design theory' espoused in *Of Pandas and People*."[77] Buchanan, for his part, liked intelligent design and valued it for its differences from creationism. It was, he argued, not biblically based but philosophical and scientific. "If intelligent design is creationism or fundamentalism in drag," he wrote, then why did "Aristotle, who died 300 years before Christ, [conclude] that the physical universe points directly to an unmoved First Mover?" Aristotle was no biblical literalist, and Buchanan, at least publicly, argued for the nonbiblical view of design. "While Darwinism suggests our physical universe and its operations happened by chance and accident," he concluded, "intelligent design seems to comport more with what men can observe and reason to."[78] For this reason, Buchanan defended intelligent design's entry into public school curricula across the country. He was not alone. Creationists in various states, cities, and schools began pushing for a new alternative to Darwinism, and the populist, grassroots creationists turned to intelligent design and argued that it should be taught alongside evolution.

One of the earliest skirmishes over ID and public schools came in Kansas, when in 1999 the Kansas State School Board voted, as chronicled in *The New York Times*, "to delete virtually any mention of evolution from the state's science curriculum." In covering the event, *The New York Times* drew a slight distinction between creationism and ID, observing, "Many creationists believe the Bible shows life on earth cannot be more than 10,000 years old. Some have adopted a less religious interpretation, saying the earth was created by an 'intelligent designer' because it is simply too complex to be explained any other way." Attempts to get the words "intelligent designer" in the curriculum were eventually defeated, but the Darwinless version was approved 6–4.[79]

The battle over evolution in Kansas went back and forth, and before the second bout took place in 2005, the native Kansan and historian Thomas Frank attended one of the Kansas anti-evolutionary conferences. "Politics," he wrote, "is when the people in the small towns look around at what Wal-Mart and ConAgra have wrought and decide to enlist in the crusade against Charles Darwin."[80] In Kansas, Frank was stunned by the mixture of prosaic, intellectual presentations by ID advocates and the entertaining showmanship of creationists. While the creationist audience snoozed during a presentation from "an Intelligent Design theorist who lectured monotonously on the faked evidence supposedly used by evolutionists," the crowd returned to life when "The Mutations, 'three fine Christian ladies' in pink dresses who strutted and whirled like an early-sixties girl group[,] . . . proceed[ed] to sing 'Overwhelming Evidence,' a ditty set to the pulsing beat of 'Ain't No Mountain High Enough.'" For Frank, this admixture of intellectual anti-evolution and populist entertainment was difficult to comprehend, but he realized that while "the audience had plainly been bored by the preceding recitation of science's errors," the "lighthearted bit of persecuto-tainment hit exactly the right note."[81]

The mixture of ID's more elitist and intellectual conservatism and creationism's more populist political persuasion, as well as its skepticism of technocratic experts and loathsome G-men, was a potent one—politically. As George Marsden noted of the *Scopes* years, anti-evolution was one of the fundamentalists' best methods of recruitment.[82] Its power in this fashion is one thing that has not changed in the intervening century. Scientifically, though, the differences were more difficult to iron out. What happens when anti-evolution is all conservatives can agree on? The enemy of my enemy may be my friend, but if ID was truly able to build a "big tent" and create a wide, but shallow, political coalition, it would be difficult to survive if its political success exceeded its scientific achievements. Without a serviceable scientific theory of design, the political movement to get design in schools would be a house built on sand.

Politics Outruns Science

The quick entry of design into public schools caused William Dembski to reflect on the precarity of ID's situation. In 2004, he addressed concerns about whether ID's successful political momentum was potentially dangerous. For one thing, he objected to the (implicitly neoconservative) idea that ID held only pragmatic value in "defeating ideologies that suffocate the human spirit." Unless it produced a real, viable scientific research program, this "noble lie" approach would be "insupportable." In addition to vexation about ID's instrumental political function, Dembski pondered whether enthusiasm for design would outrun its research program's ability. The scientific component of ID needed to be bedrock, not its politics. He noted concerns expressed by others that ID was "being hijacked as part of a larger cultural and political movement. In particular, intelligent design is thought to have been prematurely drawn into discussions of public science education." Dembski admitted that there was a "kernel" of truth in this fear and granted that ID "does need to succeed as a scientific enterprise to succeed as a cultural and political enterprise. In other words, the instrumental good of intelligent design cannot race ahead untethered from the progress of its intrinsic good." Nevertheless, even though the science of ID was logically prior to its cultural and political campaign, that did not mean science was temporally prior. It could be necessary for cultural wins to precede scientific ones, since "science grows within a cultural matrix but at the same time shapes that matrix." Altogether, though, "the scientific research and cultural renewal aspects of intelligent design need to work together, protecting and reinforcing each other."[83] It would be much harder to succeed if one were given total priority. If ID won on culture but did not have the scientific paradigm to back it up, then the result could be disastrous. Around the country, however, excitement for ID brewed, and it broke into schools not just in Kansas but everywhere, regardless of whether the scientific research program was ready.

In 2002, the Ohio State Board mandated that high school biology teachers "critically analyze" evolutionary theory.[84] Also in 2002, the Cobb County Board of Education, in Georgia, approved a sticker to be affixed to its science textbooks that read, "This textbook contains material on evolution. Evolution is a theory, not a fact, regarding the origin of living things. This material should be approached with an open mind, studied carefully, and critically considered."[85] The sticker was the basis for an eventual lawsuit that would become *Selman v. Cobb County School District*. The battle continued to oscillate in Kansas, too; the school board members who vetoed evolution were voted out of office, and the theory was reinstated, only for a new Kansas board to remake the science standards again in 2005 along ID lines—including openness to supernatural explanations and a declaration that evolution was a theory in crisis and that ID was a viable alternative. These adjustments would lead directly to the Kansas evolution hearings in 2005.[86]

Political and legal attempts to undercut evolution in US science education were not confined to the county or state level. In 2001, Senator Rick Santorum of Pennsylvania took skepticism of evolution to the federal echelon when he sponsored the "Santorum Amendment" to the sweeping education reform legislation known as No Child Left Behind. With the assistance of Phillip Johnson, who took credit for the amendment's drafting, Santorum addressed Congress in order to persuade the upper chamber of its modest goals. It was the purpose of science education in public schools, he quoted from the amendment, to "prepare students to distinguish the data or testable theories of science from philosophical or religious claims that are made in the name of science." Commenting on the effects of the amendment, Santorum elaborated, "It simply says there are disagreements in scientific theories out there that are continually tested." "I frankly don't see any downside to this discussion," he added.[87] Intelligent design went unmentioned, but the amendment was saturated with ID's characteristic political rhetoric regarding the wedge, naturalism, and philosophical bias.

The Santorum Amendment reflected the shift in the Discovery Institute's strategy toward the "teach the controversy" approach to education, in which evolution was taught more—not less. The Discovery Institute released a how-to guide for teaching evolution in 2004, in which a wide variety of materials would assist in "improving the teaching of evolution in your local schools," which would mean drawing attention to the perceived anomalies in the theory: "macro and micro" evolution, the Cambrian explosion, and Jonathan Wells's "icons of evolution," among others.[88] Scientists everywhere responded with forceful denunciations, arguing that the amendment mischaracterized science and implied that evolution was a controversial theory "in crisis," when it actually enjoyed scientific unanimity. One might argue that it was unfair to teach any scientific theory this way—after all, should gravity be portrayed in schools as a theory "in crisis" because there is not yet a unification of general relativity with quantum mechanics? (This was the angle that the satirical newspaper *The Onion* took, when it featured a story set in Kansas schools wherein the theory of gravity would be replaced with "intelligent falling").[89] Nevertheless, the benefit of this strategy was that it allowed ID to continue to press the attack against evolution without necessarily needing to present an alternative quite yet.

Senator Sam Brownback of Kansas rose to defend the Santorum Amendment and the actions of the Kansas School Board from the Senate floor, saying, "I would like to take the opportunity of this amendment to clear the record about the controversy in Kansas." Expressing frustration at the portrayal of the Kansas School Board's actions as "theocratic," Brownback (who would later serve as Kansas governor from 2011 to 2018) continued, "The school board did not ban the teaching of evolution. They did not forbid the mention of Darwin in the classroom. They didn't even remove all mention of evolution from the State assessment test. Rather, the school board voted against including questions on macro-evolution." They kept, he asserted, questions on micro-evolution. Brownback argued that this was a relatively minor move, and it only meant that the board was "trying to decipher between scientific fact and scientific assumption."[90] Separately, Brownback invoked the spirit of Kansas and compared the anti-evolutionary controversy to the abolitionist John Brown (1800–1859).[91] Despite Brownback's and Santorum's defense, the amendment was weakened while in committee, and even that form did not make it into the final No Child Left Behind bill. The National Center for Science Education, however, tried to cut through the media attention and remind that the apparent political victory of anti-evolutionists was a mirage.[92]

Despite the setback dealt to the Santorum Amendment, it appeared that evolution was being challenged everywhere. Lawsuits were filed in Georgia and Ohio. Dramatic hearings were held in Kansas regarding the scientific nature of intelligent design, hearings that were boycotted by scientists in protest. But the most lasting and determinative actions were taken in Dover, Pennsylvania, where the Dover Area School Board changed its science curriculum, announcing in a November 19, 2004, press release, "Because Darwin's Theory is a theory, it continues to be tested as new evidence is discovered. The Theory is not a fact. Gaps in the Theory exist for which there is no evidence. A theory is defined as a well-tested explanation that unifies a broad range of observations." Furthermore, the Dover board intended to draw students' attention to intelligent design, "an explanation of the origin of life that differs from Darwin's view."[93] The conflict went national when several parents of Dover High School students filed a lawsuit against the school board with the assistance of the American Civil Liberties Union and lawyers from the Philadelphia law firm Pepper Hamilton. *Kitzmiller, et al., v. Dover Area School Board* would prove to be the enduring image of the battle over evolution and intelligent design, demonstrating how ID's Icarian political success carried it too close to the judicial sun.

Conclusion

"Today in public schools across America," wrote Charles Colson (1931–2012), "the idea that science provides a fully naturalistic explanation of the world and that faith is merely a matter of religion . . . which must be kept out of the classroom, is absolutely entrenched." For Colson, the modern predicament of faith—that it had been completely privatized and walled off behind an enclosure of "personal belief"—was corrosive to Christian values. Intelligent design offered a way to change that. ID "effectively challeng[ed] this whole way of thinking. It has assaulted naturalistic evolution with lucid arguments and clear evidences of design."[94] Though ID was focused around a critique of evolutionary biology, its goals—as Colson admitted—were much higher. ID's roots were in conservative politics and the culture war, though a different strain of conservative politics than many of IDers' creationist brethren. Intelligent design's promise was to rip apart the naturalistic metaphysics that privatized religious thought, maintained the wall of separation that kept religious teaching out of schools, and sidelined religious ideas from the public square. From religious conservatives at *First Things* to cantankerous agnostics at *Commentary* to dissenting scientists like Michael Behe, ID offered a path forward. The different strains of American conservatives frequently disagreed, save for broad objection to Darwin (and even that antipathy was caused by different beliefs), but ID could be a coalition. Though ID managed to achieve some political and cultural momentum across different species of conservatism—elitist and traditional, neoconservative and instrumental, religious and populist—it was as science that it would end up judged. With the political movement gaining steam, it was perhaps inevitable that ID would find its way into the courtroom setting of the culture wars, where all such battles over the United States' litigious heart conclude. Education, and the role of science in schools and the soul of America, would be the next phase of the battle. ID proponents would tread the same ground that creationists of yore had trod from *Scopes* to *Edwards*. And so, just as it was with the creationists, the courtroom would be the site of intelligent design's greatest defeat, in the fall of 2005 in the small town of Dover, Pennsylvania.

4

Backlash

The Apex of Design

The late 1990s and early 2000s were the high period of intelligent design's momentum. Jason Rosenhouse wrote that, at the time, "ID seemed to be producing one novel argument after another. . . . [Books were] written by people with serious credentials and written with far more skill than the YEC's could muster, [and] seemed to advance the discussion in original ways."[1] The popular success of ID books—particularly Johnson's *Darwin on Trial*, Jonathan Wells's *Icons of Evolution*, and Behe's *Darwin's Black Box*—flowed naturally into a public campaign. Politicians including Rick Santorum and President George W. Bush gave support to ID, voicing their desire to see it in US public schools. "Both sides ought to be properly taught," said Bush. "Part of education is to expose people to different schools of thought."[2] The renowned atheist philosopher Antony Flew announced a conversion to an "Aristotelian" theism influenced by ID's attack on theories of abiogenesis.[3] In schools, boards in Georgia, Ohio, Kansas, and Pennsylvania began to modify curriculums to include more criticism of Darwin and sometimes even feature ID as an alternative theory. It was therefore a matter of when, not if, there would be a backlash against ID mustered with equal and opposite vigor. The contest between ID and its opponents culminated in 2005 with the *Kitzmiller v. Dover* trial, a legal conflict that convulsed the nation and dealt ID a staggering defeat. Shortly on the heels of the *Dover* decision, ID faced an energetic two-pronged assault from two very different foes: the New Atheists and theistic evolutionists (particularly those at the newly founded BioLogos Foundation). For the remainder of the decade, ID had to endure a frenzied assault.

Big Trouble in Little Dover

The journalist Edward Humes, who documented the *Dover* trial extensively, remarks that ID's greatest strength was also its greatest weakness: that is, its theistic bent made it attractive to anti-evolutionists of all stripes, but this could not be admitted in court. Furthermore, ID's nebulous nature made it popular and flexible; it also, as the testimony of the Dover board showcased, made it vulnerable.[4] ID may have been a more cerebral form of anti-evolution, but it was riding the crest of a populist wave. The alliance between these different wings of anti-evolution, and the struggle for ID to cleanly separate itself from the creationists on whom it depended for social support, would play out dramatically in *Dover*.[5] In the Santorum Amendment, mirroring the rhetoric from the Discovery Institute, the focus was less on design and more on evolution—the move to "teach the controversy." In the midst of the first Kansas controversy in 1999, Johnson and Behe penned op-eds in *The Wall Street Journal* and *The New York Times* articulating the strategy, which Thomas Woodward summarized as, "instead of teaching less evolution, schools should teach 'far more about evolution.'" That is, they should teach "[about] the legitimate scientific controversy over Darwinism."[6] Science teachers, ID leaders hoped, would highlight where Darwin was weak, but Darwin should not be removed. Wrote Behe, "If we want our children to become educated citizens, we have to broaden discussion, not limit it. Teach Darwin's elegant theory. But also discuss where it has real problems accounting for the data, where data are severely limited, where scientists might be engaged in wishful thinking, and where alternative—even 'heretical'—explanations are possible."[7] But would the creationists who were underneath the ID umbrella follow this approach or act on their own?

The campaign to discredit evolution in Dover was spearheaded by the school board member William Buckingham and the board president Alan Bonsell, both of whom were young-Earth creationists. In a board meeting on June 7, 2004, when the controversy was beginning to brew, Buckingham reportedly complained that the school's textbook (authored by the Catholic biologist and later expert witness for the plaintiffs Kenneth Miller) was "laced with Darwinism." At this juncture, months before the lawsuit, Bonsell and Buckingham used explicitly creationist and religious language when calling for a reassessment of the school's texts. At a heated meeting the next week, on June 14, Buckingham made a defiant rhetorical proclamation: "Two thousand years ago, someone died on a cross. Can't someone take a stand for Him?"[8] Buckingham's language reflected the culture-war tone of the conflict—in total, the controversy over evolution was part of a broader series of issues. Buckingham, a retired police officer, prison guard, and altogether imposing figure, saw evolution as part of a continuum of dangerous ideas (in the same way as the creationists chronicled by Toumey) and was overheard saying after the June 7 meeting that the United States "wasn't founded on Muslim beliefs or evolution. . . . This country was founded on Christianity, and our students should be taught as such."[9] Likewise, he believed that the separation of church and state was nowhere in the Constitution; it was instead a concept foisted on the pious US public by liberals in "black robes." All these comments were recorded in the local newspapers *The York Dispatch* and *The York Daily Record.*[10]

A critical strategic shift took place after the story began to get national traction. Still in June, William Buckingham contacted a Discovery Institute lawyer named Seth Cooper with a plea for assistance. As Humes notes, the details of the conversation were disputed, but it resulted in Buckingham becoming emboldened enough to push for intelligent design in the school as an alternative to evolution. Cooper, from his side, maintained that he urged the board only to follow the "teach the controversy" approach and not to incorporate ID into the school's science classes. Buckingham then reached out to the Thomas More Law Center (TMLC), a conservative Catholic advocacy law firm, which suggested *Of Pandas and People* as the go-to anti-evolutionary textbook for the school, and thus the Dean Kenyon textbook occupied center stage again during ID's apex.[11] Buckingham made several attempts to wedge *Of Pandas and People* into the classroom, and the subsequent grappling over new textbooks for the biology classes became a frustrated back and forth. When the acrimonious conflict concluded with the board vetoing his decision and with the science teachers resisting the use of *Pandas*, a donation of fifty copies of the book mysteriously appeared anyway, earmarked for use in class not as a mandatory textbook but instead as an optional reference. Despite the surprise, *The York Daily Record* proclaimed this a "reasonable compromise." However, in mid-October 2004, the board voted to change the official curriculum to include instruction in "gaps/problems in Darwin's Theory" and to make students "aware . . . of other theories of evolution including, but not limited to, intelligent design." A disclaimer to be read in science class regarding evolution and intelligent design was announced in a November 19 press release.[12] On December 14, several parents of students in Dover High School filed a lawsuit against the Dover Area School Board.[13] The parents sought help from the ACLU, which then solicited further assistance from Pepper Hamilton, a major firm in Philadelphia. Two lawyers, Steve Harvey and Eric Rothschild (both religious—Harvey Catholic and Rothschild Jewish—and both theistic evolutionists), joined the case pro bono to work with the ACLU's Witold Walczak.[14] For the defense, the Thomas More Law Center—a conservative Catholic advocacy group in Michigan, which was founded and backed by the Domino's Pizza magnate Thomas Monaghan and which boasted Rick Santorum on its advisory board—took up the cause for the Dover School Board, with the Discovery Institute following uneasily in tow.[15] More important, at least from a legal perspective, was the sudden disappearance of creationist language from the mouths of any of the pro-ID school board members. In fact, during their pretrial depositions, every one denied ever having used the word "creationism" to describe the changes to the curriculum.[16]

Thus was solidified the curious coalition of the Dover board: defendants who were led largely by two young-Earth creationists (Buckingham and Bonsell) yet defended by the legal analysts from a conservative Catholic law firm (and, indeed, named after a man in Thomas More, 1478–1535, who embodied the potential for violent conflict between Protestants and Catholics) and then the more cautious and ultimately frustrated Discovery Institute, which insistently thundered that ID was not creationism. Such an alliance reflected the sociopolitical "realignment" documented by James Davison Hunter, in which the "culture wars" necessitated a shift from denominational fealty to political and cultural pacts, meaning that old enemies like evangelical Protestants and Catholics were putting aside their differences for the purpose of political expediency on social issues like abortion, gay marriage, and creationism.[17] That these groups did not have the same beliefs was mystifying to outsiders unlearned in the differences between them. Observers who came to Dover to cover the trial, like Matthew Chapman (a screenwriter and great-great-grandson of Charles Darwin), sometimes misunderstood and mischaracterized each group. At one point, Chapman mocked TMLC's lead council, Richard Thompson, for not being sufficiently committed to "biblical literalism" (failing to adhere to the law in Deuteronomy), while later impugning evangelicals for opposing birth control as a "mortal sin."[18] Chapman's inversion of these two stances (his ire about birth control would be more accurately directed toward the Catholic Church, and his mockery of biblical literalism to evangelicals) reflected a deeper confusion about the goals of these tenuously allied factions, but he did also recognize that their aims were different, even if the details were unclear. He asked, upon learning that Jonathan Wells belonged to Sun Myung Moon's Unification Church, "Was there ever a tent (revival) this ecumenical?"[19]

As each side prepared, it became clear that all was not well for the defendants—cracks in the fragile coalition showed up almost immediately. Internecine squabbles were perhaps inevitable in the "big tent" under such conditions, but the ease with which they surfaced did not bode well for the defense. Relations between the Thomas More Law Center and the Discovery Institute broke down swiftly. Two days before the suit was filed, Phillip Johnson stated, "What the Dover board did is not what I'd recommend," and expressed preference for the Santorum Amendment approach—that is, "Just teach evolution with a recognition that it's controversial. . . . A huge percentage of the American public is skeptical of it. This is a problem that education ought to address."[20] On the day of the lawsuit's filing, the Discovery Institute publicly tried to distance itself, with John G. West declaring, "Although we think discussion of intelligent design should not be prohibited, we don't think intelligent design should be required in public schools."[21] Over the course of the year, before the courtroom hearings that would take place in late 2005, several of the biggest name expert witnesses for the defense withdrew due to conflict with the TMLC, including William Dembski, Stephen Meyer, and John Angus Campbell. They wanted their own lawyers, partly due to the court's interest in their involvement with *Of Pandas and People* and the Foundation for Thought and Ethics. The TMLC refused and fired Dembski and Campbell. Meyer later withdrew. It was so late in the game that the deadline had passed to add new experts.[22]

The tensions were not confined only to those who were involved in the trial, as outside observers noticed it too when they covered the event. The science writer Gordy Slack recalled spending time with an Italian neoconservative journalist named Guilio Meotti, who was interested in the religious nature of American people while not sharing their religious beliefs. Meotti told Slack that in Italy "no one really believes anything" and that he found Americans' evangelical piety "beautiful." For Meotti, "neo-Darwinism" paved the way for "relativism, euthanasia, embryo manipulation, and genetic engineering," and he placed great hope in Americans' religious hostility to it. However, when Slack, Meotti, and Matthew Chapman went to see a screening of a Kent Hovind video about creationism that had drawn a crowd in Dover, Slack realized that Meotti had gone missing. He surmised that the Italian journalist had been unnerved by the "self-righteous celebration of ignorance [in] Hovind's performance."[23] In Slack's view, ID's subtle campaigning had been "one of the great promotional and public relations accomplishments of the twenty-first century," but the Discovery Institute itself worried that ID was "still too closely associated with Biblical creationism," and this misunderstanding could lead to its undoing. The Discovery Institute leaders, Slack opined, "must be wringing their hands like nervous parents. . . . The complex and delicate rhetorical, PR, and legal machine they've designed . . . has been turned over to Richard Thompson, a self-righteous, ideologically driven, scientifically and philosophically naïve lawyer."[24] As Humes contends, "The Discovery Institute had to pick its battles—and its bedfellows—with great care. If the Institute allied itself with the wrong spokesperson or the wrong case, the whole movement could be irreparably damaged."[25] With intelligent design's fate in the hands of Richard Thompson, Alan Bonsell, and William Buckingham, the stage was set for just such an event.

The Trial

In September 2005, as the trial was getting under way, Dembski (despite already withdrawing from events) predicted that there was only a 20 percent chance that the Dover board would prevail in the trial. All was not lost, for he reasoned that the likeliest outcome (70 percent) was that the board would be defeated but ID would be left intact as a scientific postulate. He suspected only a 10 percent chance that the Dover policy would be, along with ID, "ruled as nonscientific." He located his skepticism in the perceived limitation of the Thomas More Law Center's resources, as the ACLU was "completely outmanning" them. But there was some reason for optimism, for the presiding judge, John E. Jones III, was a Republican, a Lutheran, and a Bush appointee. Years earlier, he had run for Congress and was supported by Rick Santorum.[26] As another admin at Dembski's blog wrote in response to him, "Have more faith, Bill! . . . Judge John E. Jones . . . is a good old boy brought up through the conservative ranks. . . . Unless Judge Jones wants to cut his career off at the knees he isn't going to rule against the wishes of his political allies. Of course the ACLU will appeal. This won't be over until it gets to the Supreme Court. But now we own that too."[27] Judge Jones himself, described by *The New Yorker*'s Margaret Talbot as a cross between Robert Mitchum and William Holden, took great pains to maintain the seriousness and integrity of the courtroom.[28] Fearing a circus, he opted not to allow television cameras in the court, though he later regretted that this prevented the educational testimonies of the plaintiffs' witnesses from being broadcast as widely as they could have been.[29] A cluster of questions animated Jones's approach to the case (and because this was a bench trial, there was no jury; he would be ruling by himself). First, and relatively narrow compared to the other three, was the question of the Dover board's actions: Were they constitutional? Was it legally permissible to feature a disclaimer apparently impugning the integrity of evolutionary theory and drawing attention to "alternatives" like intelligent design? Though this question incited the case, it entailed a whole host of more complicated questions: Was intelligent design creationism? Was intelligent design science? And was intelligent design religious?

The plaintiffs sought to demonstrate all of ID's shortcomings in one fell stroke by linking it definitively to the already-defeated scientific creationism of the 1980s. As Eric Rothschild stated in the opening argument, ID "is not identical in every respect to the creation science previously addressed by the Supreme Court in *Edwards* and other courts, but in all essential aspects, it is the same." The key thrust for the plaintiffs would be that when "confronted with that inhospitable legal environment, creationists have adapted to create intelligent design, creationism with the words 'God' and 'Bible' left out." ID was nothing but "the 21st century version of creationism."[30]

On the other side, in the defense's opening argument, the TMLC's Patrick Gillen replied, "This expert testimony will confirm the defendants' judgment by showing that intelligent design theory is not creationism, . . . [that it] is not religion or inherently religious, that intelligent design theory is science. It's a theoretical argument advanced in terms of empirical evidence, technical knowledge proper to scientific and academic specialties."[31] The ACLU and Pepper Hamilton called far more expert witnesses than did the TMLC, largely due to the Discovery Institute's withdrawal. Among the most crucial were the aforementioned Kenneth Miller, the theologian John F. Haught, the philosopher of science Robert Pennock (author of *Tower of Babel*), and the philosopher Barbara Forrest (author, with Paul Gross, of *Creationism's Trojan Horse*). Each testified on the nature of science and religion, as well as the inextricable linkage of ID to creationism.

Regarding creationism, the plaintiffs mounted a sustained assault on ID, elucidating the connections between the two and dredging up embarrassing history to showcase ID's historical lineage. Miller stated bluntly, while being questioned, that "intelligent design is inherently religious and it is a form of creationism. It is a classic form of creationism known as special creationism."[32] When Robert Pennock was on the stand, he drew distinctions between the various types of creationism and defined the term as a general hostility to evolution. When asked to describe intelligent design, he said it was "a movement that attempts to unite these various factions" in order to oppose "the naturalist world view of evolution." When asked whether he believed that intelligent design was creationism—defined in this instance as "periodic intervention"—Pennock answered in the affirmative: "Yes. It's a form of creationism."[33]

Barbara Forrest's testimony proved a devasting blow for ID. Forrest, who had already trenchantly criticized ID in *Creationism's Trojan Horse*, posed a greater and different threat to ID than did most of the other experts. Unlike figures like Miller or Haught, who were sympathetic to religion, Forrest was a signer of the Humanist Manifesto III, she was a member of the ACLU, and she had trawled the depths of ID's history to find scandalous evidence of its creationist past. The defense attempted to get her barred from the case, arguing that she did not have specialized expertise relevant to the trial and that she was "little more than a conspiracy theorist and web-surfing, 'cyber-stalker' of the Discovery Institute."[34] Jones denied the request. When Forrest testified, she produced damning evidence not only with her discussion of the "Wedge Document" that had been leaked to the internet (and that implied, from her vantage, theocratic aims on behalf of the Discovery Institute) but most importantly with the draft history of the textbook *Of Pandas and People*. After months of dedicated sleuthing by both Forrest and the National Center for Science Education's Nick Matzke, the plaintiffs' lawyers had acquired the drafts for *Of Pandas and People* through subpoena of the Foundation for Thought and Ethics. When the drafts were closely analyzed, it became obvious that *Of Pandas and People* had originally been a creationist text but that the words "creation" and "creation science" had been stricken after the 1987 *Edwards* trial. Even so, the rest of the book had not been changed—so the resulting definitions of "ID" and "creationism" were identical. Even more embarrassingly, it appeared that a search-and-find tool had been used to make the change and that it had misfired once—resulting in the curious new form of anti-evolution called "cdesign proponentsists." This vestige of the evolution of creation into intelligent design was so stark that the defense could respond only with insult, and Richard Thompson attempted to disgrace Forrest in cross-examination with questions about her membership in the ACLU.[35]

In addition to showing that ID had inextricable links to creationism, the plaintiffs also argued that intelligent design was fundamentally religious. Pennock agreed that ID was a "religious proposition" because its key idea was that "the features of the natural world are produced by [a] transcendent, immaterial, non-natural being, that's by itself a supernatural, a religious proposition."[36] Haught gave a brief history of the teleological argument, stating at the outset that intelligent design was "essentially a religious proposition."[37] The paleontologist Kevin Padian testified that intelligent design was concerned with notions of teleology, "a philosophical and overtly religious notion that is absent from ideas of evolutionary biology."[38] ID was faced with a double accusation: it was religious because of both its creationist background and its natural theological heritage.

Could the design argument be reframed as primarily scientific? And just coincidentally religious? The ID defense lawyers tried to take this tack—that it was a scientific theory with potentially religious implications—but expert witnesses for the plaintiffs rebuffed them. Not only was intelligent design religious in its impulse, but it was also not science at all. Kenneth Miller drew an analogy from baseball and the recent comeback that the 2004 Boston Red Sox had managed over the hated New York Yankees, en route to their first World Series win since 1918. One might, Miller suggested, believe that the Red Sox came back from a 3–0 series deficit because "God was tired of George Steinbrenner." Miller elaborated, "In my part of the country [Rhode Island], you'd be surprised how many people think that's a perfectly reasonable explanation for what happened last year. And you know what, it might be true, but it certainly is not science."[39] Both Haught and Padian described intelligent design as a "science stopper," with Haught saying that it walled off inquiry and blunted scientific investigation of nature: "it suffocates, I think, the scientific spirit intellectually."[40] The disclaimer of design and evolution that had been presented to Dover's students, an announcement without any discussion to follow it up, bothered Padian, who testified, "when you tell students that these ideas should be considered but then you forbid discussion, you forbid questions."[41]

It was a bleak sequence for intelligent design as the expert hearings for the plaintiffs unfolded. Even when the defense lawyers managed to score some points, they seemed to be unprepared or unable to effectively pin down their opponents. For instance, TMLC's Robert Muise asked Kenneth Miller if in the "ordinary meaning of the word a creationist is simply any person who believes in an act of creation," to which Miller agreed. Then Muise asked, knowing that Miller was a Catholic, whether he was therefore a creationist, to which Miller admitted the affirmative, because "any person who is a theist, any person who accepts a supreme being, is a creationist in the ordinary meaning of the word because they believe in some sort of creation event." Despite this admission, however, and Miller's later explanation that creationism usually meant the young-Earth version (to which he did not adhere), Muise did not press the point, instead suggesting a break.[42] That the defense missed an opportunity to highlight the differences in types of creationism was not lost on critics. As Matthew Chapman ruminated after this exchange (as well as after Miller's admission that the panspermia hypothesis was at least scientifically permissible), "It was an anticlimactic ending when it could have been so great." Chapman wondered why Muise did not ask, "You're a witness for the plaintiffs but in fact you are a creationist who believes that it's scientifically possible that all life on earth came about because of aliens?"[43] Such a move would not have won the case by itself, but it would have drawn more attention to the slippage between definitions of creationism. Dembski had expressed his desire, before the case, to show that Miller and theistic evolutionists like him would "come across as the closet ID theorists they really are."[44] However, when such a moment might have arisen, the TMLC moved on to other matters.

Similarly, Richard Thompson challenged John F. Haught by asking about his deposition statement, in which he had disagreed with Robert Pennock and Barbara Forrest for conflating ID with creationism. Haught waffled in his response and reaffirmed his view that ID and creationism were different but that this did not mean they were "opposite." Thompson asked, "So, it is wrong for the Court to get an impression that creationism and intelligent design are the same thing?" Haught cautiously replied that "they're not exactly the same thing." Right at the cusp of this point, however, Thompson moved on to ask about genes.[45] Onlookers were surprised. Humes writes, "Instead of consolidating this small victory, Thompson launched an ill-conceived attack on the whole idea of common descent, which even many ID proponents accept as at least possible, but which he personally disbelieves."[46] Chapman noted that Thompson's blustering rudeness during the entire exchange—at one point quipping at Haught, "Well, you're not the legal expert, are you?"—caused an uncomfortable silence to fall on the courtroom. Chapman ventured that this did not impress Judge Jones.[47]

In the first phase of the courtroom conflict, the ID side had fared terribly—the expert witnesses for the plaintiffs had landed sustained blows and laid out embarrassing evidence. The defense had been unable to muster a significant counter, despite some opportunity to do so. However, as Humes notes, things could have improved for ID once it was the defense's turn to call expert witnesses. Could they "get the defense back into the fight?"[48] The defense's first witness was its most important: Michael Behe. Behe did well at first, explaining the concept of irreducible complexity at detailed length. His long, technical explanation of the complexity he saw in the bacterial flagellum motor, the blood clotting cascade, and the immune system proved difficult for onlookers (and even the court) to comprehend. After the long testimony concerning the flagellum, Muise stated, "Your honor, we're about to move into the blood clotting system, which is really complex." To which Jones replied, "Really? We've certainly absorbed a lot, haven't we?"[49] Despite the difficulty in following Behe, writes Humes, he was "a good witness, and a persuasive one." After his testimony, according to Humes, some critics who were dismissive of ID wondered if "maybe there was something to it after all."[50]

Yet, despite Behe's promising showing at first, he would falter under a withering cross-examination from Pepper Hamilton's Eric Rothschild. Rothschild challenged Behe on the definition of science he employed, noting that ID could not be considered scientific under the definition of science in use by the National Academy of Sciences, and he stated that Behe used a broader definition of the term "theory." Rothschild then asked if, under Behe's definition, astrology or the ether theory of light would be considered scientific, to which Behe admitted the affirmative. His definition of "theory," the ID biochemist explained, "does not include the theory being true, it means a proposition based on physical evidence to explain some facts by logical inferences."[51] Rothschild also extracted from Behe that his famed book *Darwin's Black Box* had not been rigorously peer-reviewed and that intelligent design did not propose an alternate mechanism to evolution and was therefore only a negative, not a positive, argument (contra to Behe's expert witness testimony under Robert Muise's guidance).[52] With regard to visual spectacle, the greatest blow against Behe came when Rothschild examined him on his statement that there were no peer-reviewed explanations for the gradual evolution of the immune system. Rothschild produced stacks of papers and set them all about the dock, asking Behe if "these articles rebut your assertion that scientific literature has no answers on the origin of the vertebrate immune system?" Behe responded negatively, saying, "the literature has no detailed rigorous explanations for how complex biochemical systems could arise by a random mutation and natural selection and these articles do not address that." "So these are not good enough?" asked Rothschild.[53] Gordy Slack, looking on, wrote that Behe's claim rang "hollow as a bell."[54]

The other two expert witnesses for the defense did not do much to improve the situation. Steve Fuller, a philosopher of science from

Warwick University, was not exactly helpful to the ID cause, for he repeatedly drew attention to the fact that it was too young to offer the wide-ranging explanatory proposals that evolution was able. Furthermore, he also admitted that ID was a form of creationism.[55] For Fuller, ID might one day develop into a full scientific theory, but it was being maligned and excluded from science because the system of peer review leads to fields "bottlenecked by a few people who kind of make all the decisions in effect."[56] Walczak followed up during cross-examination by asking about a variety of other scientific theories that overcame ideological opposition to achieve acceptance and asked whether ID was so unfairly targeted that "the only chance it has is for a federal judge to order that it be taught in schools?"[57] When ID's last expert witness, the microbiologist Scott Minnich, came to the stand to discuss the bacterial flagellum again, Judge Jones quipped, "We've seen that." "I kind of feel like Zsa Zsa's fifth husband," replied Minnich. "I know what to do but I just can't make it exciting."[58]

While the expert testimony was ongoing, the legal backdrop—Was it constitutionally permissible to teach ID in schools?—faded a bit as the questions of ID's historical relationship to creationism and its scientific and religious status predominated. However, the legality of teaching ID in schools would resurface and reach a crescendo with the testimony of the board members, particularly William Buckingham. Buckingham was summoned back from his new home in North Carolina, to which he had moved before the trial had even begun.[59] While on the stand, a visibly reduced Buckingham, who entered court with a cane, was confronted with explaining two things that did not add up: Why did the board members claim that they had never used the term "creationism"? And where had the *Of Pandas and People* copies come from? For the first, Buckingham relied on the old Yogi Berra defense: "I really didn't say everything I said." He repeatedly denied ever speaking the things attributed to him by the local newspapers, or he claimed that they were said at other times and incorrectly reported with the evolution controversy. He disputed their reporting on his comments about the US being founded on "Christian beliefs," his disdain for liberals in "black robes," his proclamation of Christ's crucifixion, or his belief that creationism should be taught.[60] His defense fell flat when a short clip from the local news—which was dated from June 2004 and which had been dug up by the local journalist Lauri Lebo—was shown in court. In it, Buckingham was seen on screen saying, "My opinion, it's okay to teach Darwin, but you have to balance it with something else, such as creationism."[61] Buckingham claimed that this was a mistake, that he had meant to say "intelligent design" but said "creationism" instead because of the incessant drumbeat of local news constantly repeating it. This claim did not hold up, however, for Pepper Hamilton's Steve Harvey reminded him that he had only just recently stated under oath that he had never read any of the newspaper stories about the board's decision.[62] Here the "big tent" had finally been hoisted by its own petard—the fragile alliance between creationists of all stripes melting down under the contradictions and incongruities of the pact that had been struck. Buckingham was not yet off the hook, for it also came out in the cross-examination that he knew precisely where the *Of Pandas and People* donation had come from. The money had been raised at his church, at his prodding, and was paid for with a check from Buckingham to Donald Bonsell, board president Alan Bonsell's father.[63] Buckingham attributed his memory problems to an OxyContin addiction he had developed following a workplace accident.[64] Even though Buckingham remained obstinate under cross-examination, at the end of his testimony, Steve Harvey "look[ed] at the man on the stand, . . . shook his head and said, 'In fact, it's over.'"[65]

Judgment Day

It was not until December 20, 2005, that Jones submitted his ruling on the case. As it was read by both the plaintiffs and defense attorneys, it became rapidly clear that ID had suffered a total defeat. Walczak extolled the decision as "victory beyond our wildest dreams."[66] Jones brutalized the Dover board, describing the "breathtaking inanity" of their actions.[67] He criticized Buckingham and Bonsell for lying in their depositions about the origin of the *Of Pandas and People* books, as well as their use of the word "creationism." "This mendacity," Jones opined, was "compelling evidence that Bonsell and Buckingham sought to conceal the blatantly religious purpose behind the ID policy."[68] In a trenchant blow to ID, Jones ventured further than simply ruling against the board policy as unconstitutional. He declared not only that ID was "religious" but also that it was "not science." He compared ID to creation science from the *McLean* decision in 1982, writing that it was premised on a "contrived dualism" that set up ID (or, in the past, creationism) in a zero-sum game with evolution. "ID is at bottom," he wrote, "premised upon a false dichotomy, namely, that to the extent that evolutionary theory is discredited, ID is confirmed."[69] Furthermore, ID made a basic philosophical error in its "bedrock assumption"—that is, that "evolutionary theory is antithetical to a belief in the existence of a supreme being and to religion in general." Altogether, the board's policy had violated the Establishment Clause, but Jones's ruling went on to declare that ID could not be considered science and "moreover that ID cannot uncouple itself from its creationist, and thus religious, antecedents."[70]

The decision was a rout. The "big tent," which had granted ID strength, was its legal undoing, for the affiliation with creationists proved fatal to its success in schools, in the courts, and in the public eye. Jones himself remarked as much, writing that ID "openly welcomes adherents to creationism into its 'Big Tent.'" The only tactic the ID side had used to distinguish it from creationism was an attempt to reduce the definition of creationism to the young-Earth variety, but for Jones, this was "only one form of creationism" and ID satisfied a broader definition of it.[71]

Denunciation from ID sympathizers was fierce. An ID rebuttal was published shortly after the case, *Traipsing into Evolution*, in which four Discovery Institute writers contended that Jones had overstepped his authority in ruling beyond the simple constitutionality of the Dover board policy and argued that Jones had "repeatedly misrepresented both the facts and the law in his opinion, sometimes egregiously."[72] John G. West responded only ninety minutes after the decision's release that it was "an attempt by an activist federal judge to stop the spread of a scientific idea . . . and it won't work."[73] Jones had anticipated this, writing at the end of his opinion, "Those who disagree with our holding will likely mark it as the product of an activist judge. If so, they will have erred as this is manifestly not an activist Court."[74] Crucially, the Dover board members behind the policy were all voted out of office (or were not up for reelection) right before the decision was levied, so Jones's ruling could not be appealed because the new board wanted nothing to do with continuing the fight.[75] It was a total defeat for ID.

Would things have gone differently for ID were it not for a few key hurdles that could not be overcome? For one, what would have happened if the TMLC had not forced a case where the Discovery Institute was hesitant? What would have happened if the school board had followed the "teach the controversy" approach instead of offering *Pandas* as an optional alternative? The way the case evolved was the worst-case scenario for ID: dragged into a battle by belligerently activist creationists and then publicly humiliated when their own experts struggled to articulate what exactly ID was. Ultimately, the most damning admission might have come from Behe's mouth himself, when he confessed on the stand that the "plausibility of the argument for ID depends upon the extent to which one believes in the existence of God."[76] At root, Behe's admission showed that there was something fundamentally nonempirical at the heart of ID, that it was based on philosophical assumptions, on metaphysics, and on first principles. It therefore depended on a prior commitment. ID advocates had long claimed that evolution was spurious science for this exact reason, but after *Dover*, the shoe found itself on the other foot. Moreover, this question would also be the grounds for a new attack on ID, from a vituperative and politicized enemy that rose to meet it in public battle and take the fight directly to the heart of the matter: God's existence.

New Challenger Approaching! The New Atheism

"I am attacking God," wrote Richard Dawkins in his 2006 surprise bestseller *The God Delusion*, "all gods, anything and everything supernatural, wherever and whenever they have been or will be invented."[77] Dawkins played an unwittingly pivotal role in the founding of intelligent design: it was his book *The Blind Watchmaker* that Johnson had first read during his sabbatical in England and that had stoked the lawyer's fire as he compared it with Denton's *Evolution: A Theory in Crisis*. This was not just coincidence, either. Since Dawkins's most well-known contribution to science, *The Selfish Gene*, he began to publish increasingly on topics related to religion and science, including *The Blind Watchmaker*, *River out of Eden*, and *A Devil's Chaplain*. In 1995, the software architect Charles Simonyi endowed a chair for Dawkins at Oxford University in the "Public Understanding of Science," a position from which its holder would "make important contributions to the public understanding of some scientific field" and "interact with political, religious, and other societal forces, but they must not, under any circumstances, let these forces affect the scientific validity of what they say."[78] From this position, Dawkins promoted a thoroughly naturalistic and atheistic approach to science, animated by a singular conviction that religion was science's (and humanity's) greatest enemy. As mentioned earlier, Dawkins's conception of science-as-atheism was so central to Phillip Johnson's rhetoric that the godfather of ID frequently referred to the worldview he opposed as "the blind watchmaker thesis" and argued that it was "the most important claim of evolutionary biology."[79] As ID rose to prominence in the decade during which Dawkins occupied the Simonyi chair, it was a sure bet that he would reply in kind to the movement that he helped inspire. *The God Delusion* resulted—the rare book without a subtitle whose appellation was its thesis.

Even Dawkins and his publisher (Mariner) were caught off guard by *The God Delusion*'s runaway success. "We could barely keep the book in print," publisher Sally Gaminara expressed; "the book had struck a vital chord with the public."[80] Nor was Dawkins alone, for his tome appeared amid a great surge of atheist literature, starting with Sam Harris's *The End of Faith* (2004) and spanning the 2006 release of Daniel Dennett's *Breaking the Spell* and *God Is Not Great*, by Christopher Hitchens (1949–2011), in 2007. The explosive popularity of these works was unprecedented. As the sociologist Stephen Bullivant noted, there has been no shortage of atheist books and pamphlets in the past two centuries, but "prior to *The End of Faith*, however, none of these have sold in great numbers."[81] Together, these four men came to be known with the grandiose title of "The Four Horsemen," but the movement was broader than its central figures and included prominent scientists like P. Z. Myers, as well as the politician and author Ayaan Hirsi Ali (who in 2023 abandoned atheism and converted to Christianity).[82] The coalition was given a name by *Wired* magazine: the "New Atheism"—"New" because it was aggressive, evangelical, and intolerant of moderation. "The New Atheists will not let us off the hook simply because we are not doctrinaire believers," wrote Gary Wolf in the *Wired* article; "they condemn not just belief in God but respect for belief in God. Religion is not only wrong; it's evil. Now that the battle has been joined, there's no excuse for shirking."[83] The New Atheists were not here to make friends and converts with an irenic tone.

Much ink has been spilled to discern not only what this loose affiliation of atheist polemicists was but where it came from and why it was so popular.[84] Altogether, there are two proximate causes that bear the most significance. First, it is impossible to divorce the phenomenon from the September 11, 2001, attacks. In fact, Dawkins penned a *Guardian* article only four days after the attacks titled "Religion's Misguided Missiles," in which he argued that there was one root cause for such staggering violence. "I am trying to call attention," he wrote, "to the elephant in the room that everybody is too polite—or too devout—to notice: religion, and specifically the devaluing effect that religion has on human life." The very existence of religion posed a threat to civilization. "To fill a world with religion, or religions of the Abrahamic kind," Dawkins concluded, "is like littering the streets with loaded guns. Do not be surprised if they are used."[85] A few days later, Dawkins wrote that it was time for atheists "to get angry." While admitting that theology was not usually the impetus for terrorist attacks—whether in New York or Belfast—it nevertheless makes violence possible by otherizing groups of people as a disposable "them." "Things are different after September 11th," he wrote; "let's stop being so damned respectful!"[86] Islamic terrorism was one of the New Atheists' specific targets, but they rarely distinguished between the beliefs of radical outliers and the doctrinal content of Islam (and in fact sometimes denied that such a distinction existed). Sam Harris and Ayaan Hirsi Ali were the most conspicuous in this regard. Harris's *The End of Faith* featured the chapter "The Problem with Islam," which contained the declaration, "we are at war with Islam." He clarified that he did not mean simply extremists or terrorists but with Islam in every facet, writing, "We are at war with precisely the vision of life that is prescribed to all Muslims in the Koran." Ayaan Hirsi Ali wrote often, in her view, of the incompatibility of Islam with liberal democratic politics and documented her own life as a Muslim convert to atheism.[87] Dawkins again echoed such comments in *The God Delusion*: "As long as we accept the principle that religious faith must be respected simply because it is religious faith, it is hard to withhold respect from the faith of Osama bin Laden and the suicide bombers. The alternative, one so transparent that it should need no urging, is to abandon the principle of automatic respect for religious faith. This is one reason why I do everything in my power to warn people against faith itself, not just against so-called 'extremist' faith. The teachings of 'moderate' religion, though not extremist in themselves, are an open invitation to extremism."[88] This attitude extended beyond Islam to Christianity as well, and the threatening figures whom the New Atheists perceived within Christianity—extremists who were poised to undo the infrastructure of modern liberal society—were often creationists and intelligent design advocates. Anti-evolutionists, then, were the second cause of the New Atheism's rise.

The danger of intelligent design was a common theme in New Atheist literature. The physicist Victor Stenger (1935–2014), who chronicled a history of sorts of the New Atheism, wrote, "The intelligent design movement in biology is dying a natural death. . . . [And] intelligent design in cosmology deserves to die. I am tired of shooting so many arrows into it."[89] Dawkins admitted that the argument from design was appealing at first glance but then thankfully gushed, "There has probably never been a more devastating rout of popular belief by clever reasoning than Charles Darwin's destruction of the argument from design." He linked ID with creationism and wrote that, when it came to arguments from improbability, "it doesn't make any difference if the creationist chooses to masquerade in the politically expedient fancy dress of 'intelligent design.'"[90] Elsewhere, Dawkins called creationists and intelligent design advocates (though he rarely distinguished between them) "history-deniers," to be disdained on the level of those who deny the Holocaust.[91] Christopher Hitchens repeatedly inveighed against intelligent design in *God Is Not Great*, usually taking it aside for a dressing down whenever the topic warranted it—whether he was writing about sex, the eye, or the Cambrian explosion.[92] Sam Harris described ID as a new threat but one that was "nothing more than a program of political and religious advocacy masquerading as science."[93] Daniel Dennett flatly dismissed ID proponents, writing, "They have all been carefully and patiently rebutted by conscientious scientists who have taken the trouble to penetrate their smoke screens of propaganda and expose both their shoddy arguments and their apparently deliberate misrepresentations and evasions."[94] Despite these dismissals—the fiery tone and the angry denunciations without much substantive engagement—the New Atheism's relation to intelligent design went far deeper, for they were both animated by a common understanding of the world, of science, and the way that God might be proven or disproven.

God is a hypothesis—a scientific postulate about the nature of the physical world. This was the fundamental point of agreement between ID and the New Atheism. One should examine the natural world and see what one finds: a confirmation or disconfirmation of an acting intelligence. For ID, if God really existed and really acted in the world, then such action would be empirically detectable, provided that one shed a commitment to naturalism before investigating.[95] As mentioned earlier, Phillip Johnson's alternative to naturalism was theistic realism. "Theistic realists," he wrote, "expect this 'fact' of creation to have empirical, observable consequences that are different from the consequences one would observe if the universe were the product of nonrational causes."[96] Stephen Meyer argued that the "God hypothesis" was making a comeback, in large part due to the strength of design arguments, whether they revolved around the apparent "fine-tuning" of the cosmos or design in biology. Meyer cited the historian of science Frederic Burnham's belief that "the God hypothesis is now a more respectable hypothesis than at any time in the last one hundred years."[97] Meyer followed this up with a major book, 2021's fittingly titled *The Return of the God Hypothesis*.

The New Atheists agreed that God was a scientific hypothesis, but they disagreed with ID's conclusions. For them, the God hypothesis was a disproven conjecture: a hypothesis of which—in the reported words of Pierre-Simon Laplace (1749–1827)—they "had no need." For Dawkins, the "God Hypothesis" was his primary target. He defined it this way: "There exists a superhuman, supernatural intelligence who deliberately designed and created the universe and everything in it, including us." After contributing his own "Ultimate Boeing 747" argument for "why there almost certainly is no God," Dawkins concluded that the "God Hypothesis"—which was for him the "the factual premise of religion"—was "untenable."[98] Victor Stenger argued in his book *God—The Failed Hypothesis* that science had advanced to a point at which scientists could "make a definitive statement on the existence or nonexistence of a God having the attributes that are traditionally associated with the Judeo-Christian-Islamic God."[99] Stenger even criticized the opinion levied by Judge Jones in *Dover*, disagreeing with his conclusion that ID was "not science." Stenger thought that "intelligent design is science"—but simply "wrong science"—and he worried (for the same reasons that the philosopher Larry Laudan worried about the results of the 1982 *McLean* case) that ID would regroup unless its postulates were convincingly disproven by empirical analysis.[100] ID and the New Atheism agreed on the premise; they disagreed on the conclusion.

This baseline agreement meant that some ID advocates were supportive and even welcoming of the New Atheism. "What I like about atheists," wrote Phillip Johnson, "is that although they tend to give the wrong answers, they also tend to raise the right questions." The explosive popularity of the New Atheists impressed on Johnson the hope that "the new atheist books are likely to have a healthy effect on the position of religion within the university."[101] Finally, God was being debated again in the public square. Stephen Meyer also acknowledged that ID and the New Atheism did have a similar outlook on things. "We have," he wrote, "focused our explanatory efforts on the exact same phenomenon of interest."[102] Philosophically, Dawkins was on the right track (even if he drew the wrong conclusions). Dawkins, Meyer argued, made extensive use of the same inference to the best explanation that ID found favorable, and he also showed that behind the scenes, many scientists truly were committed to metaphysical naturalism. He demonstrated, wrote Meyer, "that a metaphysical hypothesis, just as much as a scientific one, can be evaluated by evaluating whether the evidence we observe matches what we would logically expect if the hypothesis were true."[103] William Dembski even wrote an email to Richard Dawkins, telling him he was "one of God's greatest gifts to the intelligent-design movement" and to "keep at it!"—a note Dawkins documented in *The God Delusion* with considerable exasperation.[104] With enemies like this, who needs friends?

Theistic evolutionists have long commented on the essential similarity between ID and the New Atheism, seeing them as inversions of each other. The philosopher of science Philip Clayton saw ID and New Atheism as embodiments of the "conflict thesis"—that religion and science were fundamentally and necessarily at odds. Indeed, New Atheists were standard-bearers for this old idea as it was articulated by hoary figures such as Andrew Dickson White (1832–1918) and John William Draper (1811–1882)—this despite, or perhaps because of, the fact there were no historians of science in the movement.[105] Ian Barbour, in his classic *Religion and Science*, listed both Dawkins and Johnson as paradigmatic figures in the "conflict" model.[106] The theologian Conor Cunningham was even more severe, writing that "Intelligent Design seems to form a diabolic union with Dawkins's view of religion as something verifiable. In this way, advocates of Intelligent Design are the true religious heirs to Dawkins's Doubting Thomas."[107] John Evans notes that some of the New Atheists were seemingly motivated by a much deeper moral sensibility—a horror at the violence perpetuated by religious extremists, yes, but also a preference for the ethics of liberal modernity—and that, for all their talk about science, this moral foundation was probably more important (just as it often is for creationists).[108]

The New Atheism lost most of its luster in the decade after its apex, and by the end of the 2010s, its respectability had cratered. The theologian David Bentley Hart describes it as "simply one of those occasional and inexplicable marketing vogues that inevitably go the way of pet rocks, disco, prime-time soaps, and *The Bridges of Madison County*."[109] The psychiatrist Scott Alexander theorized that the New Atheism was a "failed hamartiology" (that is, a study of sin) that attempted—in a desperate bid to understand the chaos and violence of the modern world—to diagnose religion as the original sin of the human race. In his view, its proponents moved on to more political avenues, some in support of more left-leaning identity politics, though with a significant collection embracing the rhetoric of the political right.[110]

In 2019, the biologist P. Z. Myers, a once prominent proponent, proclaimed the New Atheist movement dead, writing that though he had once held great hope in it, "the credibility of science was stolen to bolster rationalizing prior bigotries." He lamented how many people "were drawn into the Church of the New Atheism by Islamophobia" and failed to learn about "the unity of humanity" as they should have.[111] The philosopher and biologist Massimo Pigliucci came to believe that the aggressive tone of the New Atheists did more harm than good, something he learned in person when an irenic and winsome tone proved effective in debating the creationist Duane Gish. "I hadn't made an impression on these folks by way of my astute science-based arguments," he recollected, "but simply by showing up and behaving like a decent human being rather than the prick they expected."[112] Some of the New Atheist leading lights, most notably Sam Harris, went on to associate themselves with the so-called Intellectual Dark Web, where their ire was directed less at religious groups than on the perceived illiberalism of left-leaning academics and news outlets.[113] Journalists for English newspapers such as *The Guardian* openly mocked Richard Dawkins for his refusal to debate the Christian philosopher and theologian William Lane Craig and later wondered if he was "destroying his reputation."[114] Over the course of the decade, the New Atheists managed to alienate nearly everyone—even those, as Alexander observed, who would ordinarily be expected to support them. For all New Atheism's bluster, it turned out to be a paper tiger, though the populist movement it unleashed across the internet (notably on Reddit's r/atheism subreddit page and on 4chan) has remained a vital force.[115]

But what about those religious believers—and there were many—who accepted evolution? The Catholic philosopher Edward Feser did not include the design argument in his book *Five Proofs of the Existence of God*. Dawkins's "Ultimate 747" argument was a side issue, as the philosophical arguments that Feser was interested in had nothing to do with probability or complexity, as the design arguments often did. These were premised on the analogy of natural objects to machinery, and Feser had no interest in this. He instead favored arguments of "strict metaphysical demonstration, not (as 'design arguments' are) mere exercises in inductive or abductive reasoning."[116] Feser would not be alone. And so, one must wonder if the real victors in the battle between ID and New Atheism were the theistic evolutionists, a third party that observed it from afar. That is not to suggest that theistic evolution is suddenly triumphant in the public square but rather that theistic evolution went from being practically a nonentity (outside of the academy) to achieving a public position, and this while its antagonists declined in popular estimation. The common villain of both ID and the New Atheism, the theistic evolutionists too had their heyday in the immediate aftermath of the Dover trial and began to build popular support that, though it arguably never equaled its two interlocutors, nevertheless proved to be a consistent thorn in ID's side.

Design's Frequent Foe: Theistic Evolution

If intelligent design could boast one true enemy, it would be theistic evolution. Even those whom one would expect to have sympathy (such as Behe, who has been accused by creationists of being a theistic evolutionist himself) have expressed distaste for the idea. Behe's primary complaint was that divine creation—or even guidance of evolutionary development—was incompatible with the heart of Darwinism, which he viewed as randomness or contingency. Theistic evolutionists, in attempting to harmonize God's creative action with an ontologically random process, were "kidding themselves if they think it is compatible with Darwinism."[117] "Theistic evolution," William Dembski argued, "takes the Darwinian picture of the biological world and baptizes it. . . . When boiled down to its scientific content, however, theistic evolution is no different from atheistic evolution." Even more frustrating, for Dembski, theistic evolution implied that God was "a master of stealth who constantly eludes our best efforts to detect him empirically."[118] In the halcyon early days of ID, Michael Denton expressed the same point, writing that Darwinian evolution was not compatible with traditional, biblically based religion.[119] Phillip Johnson conjured up the old fundamentalist dichotomy and called theistic evolutionists "modernists," arguing that they offered very little: "all modernist theologians can do is to put a theistic spin on the story provided by materialism."[120] Moreover, such modernists were unduly influenced by the mainstream scientific worldview, and he lamented that many Christian academics had been "educated in the secular academies. They've learned theistic evolution. They really don't *understand* the issues. They don't know what's wrong with the scientific evidence. They just follow blindly along where the secular world is going."[121]

ID's fundamental problem with theistic evolution was the way the latter's supporters consistently tried to harmonize evolution with religion in such a way that, as ID proponents saw it, gave the game away to the atheistic evolutionists, who viewed the universe as fundamentally chancy and lacking purpose. Johnson wrote that any attempt to get an audience with naturalists would fail and that "materialists see no need to make concessions to people whom they regard as either hopelessly wooly minded or yearning to find some opportunity to sneak some element of supernatural influence into science."[122] Forrest and Gross accused Dembski of hypocrisy for at one time stating that ID advocates were "no friends" of theistic evolution and at another time arguing that there was a logical compatibility with ID and theistic evolution. "Such flat contradictions," they wrote, "are routine in Dembski's oeuvre."[123] However, the real issue for Dembski—and for ID in general—was not so much evolution per se as it was the irreligious core they saw sitting at the heart of the Darwinian version of it. Darwinian evolution was premised on the power of natural selection acting on random mutations, a natural mechanism that ID advocates read as unsupervised and purposeless, even if it was not purely "random" in a logical sense. So, while an ID advocate like Behe might have provisionally accepted something like common descent, he would only have done so based on a guided or directed version of evolution, not an unguided one. As Dembski expressed, theistic evolution was the opponent of ID (and, we might add, the New Atheism) because it posits that God's design (or nondesign) is not "accessible to our natural intellect."[124]

While the metaphysical disagreements were real, the biggest difference between ID and theistic evolution was empiricism. For Meyer, the God of the theistic evolutionists guided nature in a "completely undetectable way" and "did not actively or discernibly guide" the evolution of nature.[125] This was unacceptable, for, as Meyer understood it, theism necessarily entailed a "personal, intelligent, transcendent God who also acts within the creation."[126] If theism is true, there will be natural evidence for it. "What is surprising," he wrote, "is that even many theologians and theistic philosophers deny that scientific evidence can *support* theistic belief."[127] Behe accused Kenneth Miller of inconsistency in promoting Darwinism but also hypothesizing that God could "act in nature without being obvious." "Which is it," he asked, "God in control or not?"[128] Divine intervention is a binary game—if God is real, then it must be there (unless deism is true, perhaps, which Meyer rejects as a deity not involved in nature—something, for him, that is suspiciously similar to theistic evolution).[129] If there is no direct, physical evidence of supernatural intervention into nature, then the best explanation is that there is no God. In this, the New Atheism and ID were at one, and the theistic evolutionists were the opposition.

Theistic evolutionists curried no favor with ID during the *Dover* trial, and in fact they played a crucial role in ID's defeat and earned for themselves a reputation as cultural quislings. Even before the case, Johnson characterized Arthur Peacocke and John F. Haught as part of a "tiny segment" of religion and science scholars, academic theists benignly tolerated by atheist scientists since they would go on the offensive for evolution (a campaign they were able to support, wrote

Johnson, because of the "considerable financial resources of the John M. Templeton Foundation").[130] At the *Dover* trial, Kenneth Miller and Haught were unabashed about their support for theistic evolution, and the ACLU-partnered lawyers from Pepper Hamilton, Steve Harvey and Eric Rothschild, both practiced their respective religions (Catholicism and Judaism) and accepted evolution.[131] Judge Jones wrote that ID's fundamental error was its inability to countenance the potential harmony between evolution and religion. As mentioned earlier, Jones viewed that ID's "bedrock assumption" of the incompatibility of evolution and theism was "utterly false." "Scientific experts," he opined, "testified that the theory of evolution represents good science, is overwhelmingly accepted by the scientific community, and that it in no way conflicts with, nor does it deny, the existence of a divine creator."[132] Theistic evolution was the trump card at *Dover*, and it would continue to dog design in the years after the trial.

In 2006, Francis Collins—the world-renowned geneticist, director of the Human Genome Project, and Christian—published a book called *The Language of God*. In it, he considered religion from a scientific vantage, and vice versa, addressing questions about creation and evolution. He discussed intelligent design as a third way between creationism and atheistic evolution but doubted its viability. "The warm embrace of ID by believers," he wrote, "is completely understandable, given the way in which Darwin's theory has been portrayed by some outspoken evolutionists as demanding atheism." However, despite this, "this ship is not headed to the promised land; it is headed instead to the bottom of the ocean."[133] In design's place, Collins suggested theistic evolution but with a new coat of paint. "Perhaps a more trivial reason that theistic evolution is so little appreciated is that it has a terrible name," he wrote; "relegating one's belief in God to an adjective suggests a secondary priority." Rather than reducing God to a modifier of evolution, Collins coined a new term: "BioLogos," which expressed "the belief that God is the source of all life and that life expresses the will of God."[134] The publication of Collins's book gave theistic evolution a publicity boost, and he explained later that he received "thousands of e-mails—primarily from other Evangelicals—asking questions about how to reconcile scriptural teachings with scientific evidence." Due to the onslaught of questions, Collins "decided to gather a group of theologians and scientists to create the BioLogos Foundation in order to foster dialogue between the two sides."[135] Collins presided over the BioLogos Foundation for two years until President Barack Obama appointed him to direct the National Institutes of Health. Darrel Falk and then Deborah Haarsma succeeded Collins as presidents of BioLogos.[136]

Considering that ID and the BioLogos Foundation share a natural constituency in evangelical Christians concerned about science, it is no surprise that they have battled repeatedly since the latter's creation. The Templeton Foundation—the bogeyman for atheist scientists like Richard Dawkins and Jerry Coyne—repeatedly endowed BioLogos with considerable financial resources as it attempted, in the words of the historian of science Edward Davis, to bring "the best scholarship on science and Christianity out of the academy and onto the street."[137] Representatives of the Templeton Foundation came to later express skepticism about intelligent design's viability, despite some previous interest in the movement. Charles L. Harper Jr., senior vice president at Templeton, told *The New York Times* that Templeton initially offered some grants conferences and courses to discuss design, but when the time came to offer research grants, ID advocates never submitted any proposals. Harper said that, while some foundation officials had some tentative interest in ID, even those who were "initially intrigued . . . later grew disillusioned." "From the point of view of rigor and intellectual seriousness, the intelligent design people don't come out very well in our world of scientific review," he explained.[138] When *The Wall Street Journal* published an article in late 2005 alleging that Templeton had sympathized with ID before moving away from it, the paper was met with a quick denunciation issued from the foundation itself. A Templeton press release stated, "Any careful and factual analysis of actual events will find that the John Templeton Foundation has been in fact the chief sponsor of university courses, lectures and academic research which variously have argued against the anti-evolution 'ID' position."[139] (It must be pointed out that Templeton did grant Dembski a $100,000 Templeton Foundation Book Prize grant to write *No Free Lunch* and *Being as Communion* so the defensive posture here is a bit retroactive.)[140] Nevertheless, ID was thwarted again by theistic evolutionists in 2009, when the Vatican held a conference on religion and evolution, for which ID advocates were not invited because conference organizers "wanted an intellectually rigorous conference on science, theology and philosophy."[141] Once again, some early interest in the design argument—particularly by Cardinal Christoph Schönborn—was seemingly sidelined as time passed.[142]

All throughout, BioLogos remained a persistent foe. The Old Testament theologian John D. Currid expressed concern that BioLogos had garnered support from evangelical pastors and scholars like Peter Enns, John Walton, N. T. Wright, and Timothy Keller, writing, "The evolutionary creation movement is stronger than it has ever been and is making inroads into evangelical thought today."[143] The old-Earth creationist Fazale Rana noted that "a growing number of evangelical and conservative Christians question the traditional treatment of the biblical account of human origins and the historicity and sole progenitorship of Adam and Eve."[144] Eventually, BioLogos struck a deal with InterVarsity Press to produce a line of books promoting the theistic evolutionary perspective, titled "BioLogos Books on Science and Christianity."[145] This series was something of a symbolic victory for theistic evolution, as it was InterVarsity that published *Darwin on Trial*, as well as several other ID works over the years (like *Mere Creation*, *Darwin's Nemesis*, and *The Design Revolution*). That InterVarsity would move on to promoting BioLogos's works with such regularity less than two decades later indicates a noticeable shift in the attitude toward theistic evolution among American evangelicals; that theistic evolutionists were getting a hearing in nonacademic settings at all was a significant development.

In 2006, *Time* magazine featured a debate with Richard Dawkins and Francis Collins. *Time* in fact noted that ID was a loud voice in the room that had triggered the "unprecedented outrage" of atheist scientists but suggested that conflict was not the only mode of conversation and that "informed conciliators have recently become more vocal."[146] So it was that the debate took place between two disagreeing evolutionists, with nary an anti-Darwinist to be heard. Despite these successes in the book and grant world, Elaine Howard Ecklund has noted that while Collins was popular with Christians and non-Christians who were interested in science, very few people outside the field knew who he was.[147] That said, people who did know of him consistently described him as an influencer in perceptions of religion and science, while Richard Dawkins largely failed to move the needle in the same way (even among atheists).[148]

Theistic evolution's modest success did not go unnoticed by the ID movement. Its retaliatory efforts culminated in the 2017 publication of a mammoth tome titled *Theistic Evolution: A Scientific, Philosophical, and Theological Critique*, in which dozens of critics lacerated theistic evolution in exacting detail, finding the entire concept, in Stephen Meyer's words, "logically contradictory" and "scientifically vacuous."[149] On the philosophical side, Meyer criticized theistic evolution for its barest-minimum philosophical stance, stating that it was at bottom "little more than an *a priori* commitment to methodological naturalism."[150] Science and philosophy did not stand alone, however, for *Theistic Evolution* also included a substantial theological critique. The evangelical systematic theologian Wayne Grudem anticipated the geological timeline problems latent in a theological attack and attempted to sidestep them with a declaration that "this book is not about the age of the earth."[151] Nevertheless, for Grudem, the first three chapters of Genesis were the fulcrum of theological debate among creationists, ID advocates, and theistic evolutionists, but the consequences reached further. In a later essay, Grudem argued that theistic evolution undermined fundamental Christian doctrines such as the Fall of Man or the atonement.[152]

There is a certain doctrinal, and even political, strand of thinking in the ID polemic against theistic evolution. Not only were the more evolutionary ID supporters—like Behe or Denton—not present in *Theistic Evolution*, but the book also manifested concern about the future of Protestantism, which it viewed (much as in the early fundamentalist days) as necessarily dependent on the hermeneutical perspicacity of scripture. Despite not being a Christian himself, Steve Fuller commended the book as "an unprecedented opportunity for educated nonscientists to revisit the spirit of the Reformation by judging for themselves what they make of the evidence that seems to have led theistic evolutionists to privilege contemporary scientific authority above their own avowed faith." In fact, theistic evolution, in addition to undermining the democratic aspect of Protestant hermeneutics, also gave license to the liberal technocracy. "Its advice to the faithful," wrote Fuller, is to "keep calm" and "trust the scientific establishment," even if it means "ceding the Bible's cognitive ground." And is this not a betrayal of the message of the Reformation, which tells Christians that they are "entitled and maybe even obliged to decide on matters that impinge on the nature of their own being"?[153] John D. Currid took up the same theme, writing that Genesis "contain[s] little indication of figurative language" and that in fact the Reformation hermeneutic depends on this. Its basic premise is that "Scripture interprets Scripture," and there is "no text in the Bible that suggests that Genesis 1–3 is a figurative passage or that would counter the basic chronological/sequential structure of the account."[154] Here, one can see ID supporters become more and more creationist in tone and fundamentalist Protestant in method. The reason why, one might suspect, is that the specter of Catholicism looms behind this struggle. If the self-evident, commonsensical approach to scripture does not work—because, say, evolution might be true and Genesis might be allegorical—then one would be forced to admit that other methods than "Scripture interpreting Scripture" would be needed, perhaps something like a Catholic Magisterium or Orthodox Holy Tradition, in which patristics can come to the rescue. More than denominational affiliation is at stake, however, as Colin Reeves also noted that theistic evolutionists "resemble . . . yesteryear's liberals."[155] They sought to change Protestantism from within. And, Currid added, even some traditional evangelicals at the BioLogos Foundation have started to question the venerable doctrine of original sin, causing Currid to fret that "Pelagianism is almost an inevitable result of the denial of the historical Adam and Eve."[156] None of this is to say that ID cannot work with Catholic theology—as, indeed, it has its Catholic supporters; but the struggle between theistic evolution and ID has increasingly become rather binary, and as time has gone on, some of ID's supporters have taken on more overt Protestant and even at times creationist rhetoric in order to distinguish themselves from their evolutionary coreligionists.

As a complete tome, *Theistic Evolution* represented all the strengths of ID: when gathered together in a loose coalition, scholars from various confessional and ideological backgrounds could muster a robust critique of an ideology they disagreed with, whether it be Darwinian evolution or its Christian supporters. As was often the case with ID, however, the best defense was a good offense, for there was little discussion of how ID might plausibly assist Christians in understanding the historicity of, say, Adam and Eve, especially when the Fall's place in history was just as acute a problem for an ID advocate such as Michael Behe as it was for any theistic evolutionist at BioLogos. Moreover, its affirmation of bedrock Protestant biblical hermeneutics and its mechanical accounts of divine action meant that the "big tent" of ID was shrinking in the years after *Dover*. It is worth meditating on causation a bit longer, for it is here that the chasm between theologically conservative ID proponents and theologically conservative theistic evolutionists became most pronounced—with each accusing the other of modernism.

Doubting Thomists

The Catholic Church seemingly embarked on a "will they, won't they" flirtation with intelligent design for a time before pulling back and maintaining a real, though sometimes cool, preference for Darwinism.[157] There had long been Catholic criticisms of evolution (as we have seen with Russell Kirk or Richard Weaver), so this turn from ID was not an inevitability. It happened because, in general, there is a serious difference in perspective between intelligent design proponents and Catholic theologians—especially devotees of Thomas Aquinas—regarding concepts of creation, divine action, and causation. This broader story can be seen writ small in the career of Francis Beckwith.

A philosopher with interest in law, Beckwith is a former evangelical Christian who converted to Catholicism (or, more accurately, returned to it since he grew up in a Catholic family) and embraced Thomistic philosophy. He was originally sympathetic with many of the ID arguments and was a fellow for a time with the Discovery Institute. His career at Baylor University was jeopardized in 2003 when he testified to the Texas State Board of Education that teaching ID in public schools "would not violate the establishment clause of the U.S. Constitution as had its predecessor 'creationism' or 'creation science.'"[158] This was the subject of Beckwith's law school dissertation and book *Law, Darwinism, and Public Education*. In his own telling, Beckwith was initially excited about ID but was most concerned with the way it might unseat "a particular school of political liberalism from its cultural dominance."[159] In the ensuing controversy about his support for ID, and potential limbo at Baylor, Beckwith began to wonder how committed he was to ID as a scientific and philosophical enterprise. In the end, he took up a critical perspective informed by Thomas.

Many Christian, and especially Catholic, intellectuals did not join up with ID. Why? For Beckwith, it was because ID "teaches us the wrong lessons about the relationship between God and creation—lessons that, ironically, concede too much intellectual real estate to Dawkins and Dennett."[160] Both ID and the New Atheism embrace God as a hypothesis—whether they are for or against it. For Beckwith, seeing God as a hypothesis makes him just another causal agent in the universe.[161] This is a problem because it places God in direct opposition to modern science. The extent to which something is caused by nature is, then, the same extent to which it is not caused by God. Because ID usually takes an interventionist view of divine-nature relations, IDers dislike it when theistic evolutionists (of any stripe) conceive of God's activity in nature as something secretive.[162] But Thomists argue that one should see design (as Aristotle did and as Thomas did after him) as intrinsic to nature, an immanent teleology, rather than extrinsic to nature, impressed on inert matter by a deity existing in opposition to the mechanical world. Thomas, contends Beckwith, observed "the design . . . in the natural world [as] intrinsic to nature, arising in our minds as a consequence of our intellectual power to know the four causes of a thing. For Aquinas, design is not a scientific hypothesis for which we seek confirmation (e.g., specified or irreducible complexity), but rather, a metaphysical truth about the natural world that is practically undeniable."[163]

The question of divine action in the world is key to the ID dispute with theistic evolution and Thomism especially, and within this dispute, there is a fundamental disagreement—and possible misunderstanding—on the difference between primary and secondary causation. And an inability to see eye to eye on causation means that ID would have always struggled to court conservative Catholics (despite getting a few adherents here and there). For Meyer, the problem with theistic evolutionists was that they restricted all of God's creative activity to "secondary causes," something they "equate with the laws of nature and evolutionary mechanisms."[164] Their contraction of God's activity in the world to such an indirect mechanism meant that theistic evolution is based on a "theologically heterodox view of divine action."[165] They view God as passive and uninvolved in creation, a God that does not "intervene" and at most "created the laws of nature at the beginning of the universe" and now "constantly upholds those laws on a moment-by-moment basis." This God, for Meyer, would not be directly involved in creation and so therefore would not be able to know what would happen; perhaps humans would never have evolved to begin with, for all this God knew.[166] In fact, as Meyer wrote along with Paul Nelson, a theistic evolutionist (such as Darrel Falk, the target of their article) would, in confining God's activity to secondary causation, be making "a commitment—perhaps a distinctively Christian commitment—to methodological naturalism."[167]

This is not a new point of contention. Norman Geisler, in his proto-ID work *Origin Science*, made a similar (though not identical) move. For Geisler, a secondary cause was completely natural—say, for instance, microevolutionary changes within species. A primary cause was a direct act of God in a singularity-type event, such as the creation of the first cell or the Big Bang. Geisler characterized a primary cause as an intelligent agent acting within nature (and, in fact, a human being creating a message would be an instance of a primary cause). And when there was an unexplained origin of new information—such as in the creation of DNA—then the logical inference would be to impute its cause to a "primary" one, that is, a designing intelligence.[168]

This is not the way Thomist philosophers, or classical theists a whole, understand primary and secondary causation. Geisler's deep confusion on this issue, moreover, crops up in some areas of the later intelligent design movement. Geisler seemed to view primary causes as secondary causes that had no physical explanation. But some ID proponents, too, adopted a similar mindset. Conceiving of the universe as a kind of machine, one so artfully contrived and complex that it must have needed a designer, many saw it as having its own kind of sustaining power. C. John Collins took up arms against intrinsic causation in *Theistic Evolution*, arguing that ID needed to go beyond the arguing about general, cosmic design and focus on specific, small-scale instances of design.[169] Crucially, however, Collins saw such small-scale design as "*imposed design*."[170] But imposition implies contention, according to Thomists, even an adversarial relationship, such that the universe has its own being unto itself and is set apart from God in a kind of binary opposition. Its secondary causes are then totally independent of its primary cause (God's constant, eternal sustaining of everything). This same either/or is present in the evangelical theologian

Wayne Grudem's critique of theistic evolution. For Grudem, theistic evolution is unpalatable because it means the creatures of the Earth were not "*directly made*" (as Genesis teaches) in "*specific acts*" (as the Psalms teach).[171] Furthermore, whereas Thomas Aquinas speculated that intrinsic teleology allowed creatures to build themselves as though a ship might be imbued with powers to assemble itself, Grudem felt that the Bible could not countenance such a view. "These creatures [in the Psalms]," he wrote, "are nowhere said to be the product of 'materials that assembled themselves' (the theistic evolution view); they are specifically the *works of God's hands*."[172] The binary is present here, too. The extent to which something is caused by natural processes is the extent to which it is not "the works of God's hands."

Many classical theists view this as a modern departure from medieval philosophy and Thomistic/Aristotelian understandings of creation. In scholastic thought, primary causation is the permanent creative activity of God, who is a necessarily existent being (I Am, the Uncaused Cause, etc.) who maintains the entire cosmos in existence at every moment. Far from occasionalism, which collapses primary and secondary causation, this view of causation understands God's creative power as a permanently sustaining act, not an endless intervention. To the extent that *anything* happens at all, God is the primary cause of all things. Were God to stop sustaining the universe, it would simply cease to be. Conversely, the theologian David Bentley Hart contends that ID heralded a modern form of the Platonic demiurge, a creative but bungling craftsman who must work with the material already in existence in order to fashion creatures. As Hart writes, this modernist picture of God is "not the source of the existence of all things but rather only the Intelligent Designer and causal agent of the world of space and time, working upon materials that lie outside and below him, under the guidance of divine principles that lie outside and above him." This deity, this demiurge, is "a maker, but not a creator in the theological sense: he is an imposer of order, but not the infinite ocean of being that gives existence to all reality ex nihilo. And he is a god who made the universe 'back then,' at some specific point in time, as a discrete event within the course of cosmic events, rather than the God whose creative act is an eternal gift of being to the whole of space and time, sustaining all things in existence in every moment."[173] And, not to put too fine a point on it, "the recent Intelligent Design movement represents the demiurge's boldest adventure in some considerable time." Instead, and though he is often highly critical of Thomism (especially Thomistic views of animals and souls), Hart cites Thomas positively, writing, "according to the classical arguments, universal rational order—not just this or that particular instance of complexity—is what speaks of the divine mind."[174]

There has been some pushback by ID supporters who contest the alliance between Thomism and Darwinism. Robert Koons and Logan Paul Gage, for instance, argued that Thomist critics of ID mischaracterize ID as modernist or mechanical by relying on one-dimensional portraits of Thomas's thought. They therefore understand "neither ID nor the heart of Darwinian evolution." Because Darwinism reduces all of nature to processes that are "impersonal, non-intentional," and limited to "material and efficient causes," that means "if Darwinism is true, then Thomism must be false."[175] While critics accused ID of indulging in mechanical philosophy, Koons and Gage contended that the information-focused arguments like Meyer's and Dembski's were really nonreductionist. Just because something was machine-like does not make it an actual machine (contra Behe), and Thomas Aquinas himself "often used analogies between living things and man-made artifacts."[176] Furthermore, Thomas was not opposed to miraculous intervention in nature and viewed creation as a whole as having an extrinsic source. In reality, what ID was really searching for was not individual instances of complexity for its own sake but rather empirical evidence of teleology in nature (whether intrinsic or extrinsic).[177] Thomas, for his part, thought that God's acts could be discovered in nature.[178] This makes ID a bigger metaphysical tent than how it was presented by some of its marginal figures (such as the contributors to the *Theistic Evolution* volume, who tended to be more literalistic and Protestant), and ID could in theory accommodate "front-loaded" views of evolution or other nonreductionist approaches to science.[179] Perhaps even critics like Hart, Conor Cunningham, and Edward Feser (each of whom, though opposed to ID, nevertheless resists the mechanistic interpretations of evolution) could find some common ground, given their professed interest in systems biology, convergence, and the extended evolutionary synthesis.

Beckwith, in reply to Koons, granted that Thomists could be guilty of overemphasizing the importance of intrinsic teleology to his thought and underemphasizing God's role as an "exemplary cause," but he nevertheless maintained the same critique that this limited metaphysical clarification did not make ID "consistent with Aquinas's thought. After all . . . nature is never *by itself*." ID was still premised on the same duality of creation either by nature or by God. They still unwittingly smuggled the very scientism they contested in through the back door. While their "hearts are in the right place," concluded Beckwith, "the way in which they conceive of God's creative power and action concedes far too much to the scientism they want to defeat."[180]

William Dembski returned to the fray to show that ID was not the cut-and-dry dichotomy of either God or nature that Thomist critics alleged. In his comprehensive statement on ID, *Being as Communion*, Dembski set out to present a metaphysics of intelligent design, focusing specifically on information. In his view, the information inherent in the universe undid materialism (for it showed that information, something conceived of by a mind, was both causally and ontologically prior to matter). Privileging information meant seeing the world in a "top-down" manner, unlike the "bottom-up" approach of materialism.[181] For Dembski, ID's main thrust is that it "asks teleology to prove itself scientifically"—it is more than just a metaphysical position. Thomists have accused ID of relying entirely on extrinsic, purely mechanistic design arguments, but ID was not limited to that view of teleology alone. It is compatible with intrinsic design too, and ID's opponents should be charitable in recognizing that ID can countenance multiple perspectives on causation and divine action. "Design," he wrote, "includes among its recognized meanings pattern, arrangement, or form, and thus can be a synonym for information. Moreover . . . intelligence therefore need not merely refer to conscious personal intelligent agents like us, but can also refer to teleology quite generally."[182] Nevertheless, in a footnote on this topic, Dembski did concede, "in fairness to Feser and [Stephen] Talbott, their criticisms are understandable because intelligent design advocates, myself included, haven't always been as clear as we might in our use of design terminology, not clearly distinguishing between external design from intelligence or teleology more generally."[183]

Edward Feser appreciated the gesture but maintained critical distance from ID, which he felt was still too wedded to this modernist mechanistic metaphysics, despite the occasional ID supporter who was not so preoccupied with extrinsic design. He agreed with ID that science's methodological naturalism evolved into a metaphysical naturalism, and the practical exclusion of telos became a belief that teleology did not exist (at least, apart from the restricted, inadvertent way characteristic of Darwinian evolution).[184] Even so, he felt that the kind of "top-down" understanding of order that moderns favored for divine–nature relations meant that matter became "entirely passive" and inert.[185] Regarding Thomas, Feser granted that he saw some things as "machine-like" but rejected the philosophy that might build itself from an analogy. Nature is not a machine or an artifact such that extrinsic design could be easily deduced from it; mechanism bans telos from reality. Because only extrinsic, Paleyite interventions in nature can fit into a machine world (as Lewis Mumford pointed out too), when Paley was disposed of, so was the design argument.[186] Though some ID proponents, like Dembski, are resistant to the idea that nature is a machine, Feser felt that Dembski did not cleanly separate himself from the mechanical understandings of nature inherent in modern science and materialist metaphysics. For Feser, a broad, general view of intelligent design (perhaps one that fits with intrinsic teleology and a "front-loaded" view of design) could maybe fit with Thomistic philosophy, but the lion's share of what constitutes ID thought could not coexist with Thomas and Aristotle.[187] Even in Dembski, when one gets down to it, the information in living things "is not 'internal' to them but must be 'imposed' from 'outside.'" After all, Dembski himself committed the cardinal Thomist sin when he said that "there are problems with Aristotle's theory, and it needed to be replaced." As Feser reads it, "Dembski rejects the Aristotelian notion that teleology is *intrinsic* to natural substances and processes in favor of the Platonic-Paleyan conception of teleology as *extrinsic* or imposed on natural substances from outside by a designer. Again, that is precisely a mechanistic conception of causality, even if it is a theistic rather than atheistic version of mechanism."[188] Dembski, Feser contended, was looking for design in nature's "deviation" from expectation, which is the last thing an Aristotelian Thomist should do; on the contrary, it is the regularity and expectation of nature that provides the strongest proof.[189] In the end, while Dembski might have proclaimed a nonmechanical view of reality and an openness to intrinsic teleology, this was more smoke than fire. Feser concluded that ID was too slippery in its terminology, making design a synonym for information and for intelligence, and that Dembski used "the term 'information' . . . in several different senses, freely sliding from one to another without always making it clear which one is supposed to be doing the work in a given argument." Ultimately, he surmised, "Dembski seems intent on sidestepping potential objections to ID by making its basic commitments as flexible as possible." But verbal compatibility does not entail substance. "Given the enormous influence of Dembski's ideas within the ID movement," Feser concluded, "no one should be surprised that Aristotelian-Thomistic philosophers have often been very critical of that movement."[190]

The point in descending into these arcane topics regarding causation, divine action, and Thomism is not to adjudicate who makes a better argument. Rather, it is to show that the divergence of views between theistic evolutionists and intelligent design advocates—both of whom emphatically reject materialism—are deep enough and substantial enough to preclude an alliance. And it does seem that ID needed the support of theistic evolutionists in order to have a better chance of success in academia. Instead of support, however, theistic evolutionists helped defeat ID at *Dover*; began publishing a large volume of books attacking ID, even doing so with the assistance of evangelical publishers; persuaded the Templeton Foundation to keep distant; and stymied ID's influence in the Catholic Church. They proved, then, to be far more dangerous opponents for ID than did the New Atheism—a movement that, in the end, shared many of ID's assumptions. Dembski made this same observation, writing in response to Howard Van Till that "theistic evolution remains intelligent design's most implacable foe."[191] That Van Till, years later, ended up leaving theism behind and embracing a "comprehensive but non-reductive naturalism" (one that accepted God as a logical possibility but not one for which he finds "sufficient justification") would probably be seen by ID advocates as proof of theistic evolution's seductive threat.[192]

Conclusion

Theistic Evolution is instructive as an especially acute illumination of ID's successes and failures. When ID proponents restricted themselves to attacking evolutionary theory (whether it was the problem of abiogenesis or the sufficiency of natural selection to explain all life's variation) or theistic evolutionists (who struggled to harmonize the historicity of the Fall, and the problem of evil, with evolutionary history), their case was extensive and appeared impressive. When on the attack, ID and its broad, big tent coalition could land blows on its opponents. However, when ID was forced to stand its own ground and articulate a positive vision of itself, and what a worldview based on design would look like, it did not succeed. At *Dover*, ID was roped into a conflict by the creationist wing of its big tent, and it wilted under the spotlight—because it could not offer a positive scientific alternative but also because it could not cleanly separate itself from its creationist allies and forebears. After the trial, Barbara Forrest and Paul Gross updated *Creationism's Trojan Horse* with a chapter on the trial and a 2004 statement from the ID advocate and creationist Paul A. Nelson, in which he admitted that "easily the biggest challenge facing the ID community is to develop a full-fledged theory of biological design." Nevertheless, he admitted, "We don't have such a theory right now, and that's a real problem." Forrest and Gross concluded, "Nelson's closing quip—'We ain't seen nothing yet'—should be taken literally. We ain't seen it and they ain't shown it."[193] In the post-*Dover* world, ID—and the Discovery Institute behind it—would divert more resources to articulating a positive vision of design and would attempt to establish a research program based on it. Nevertheless, as *Theistic Evolution* showed, intelligent design continued to operate primarily as an offensively minded movement that was never able to make its disparate positions fully cohere.

That said, it would be a mistake to assume, as most critics did, that ID disappeared or died after *Dover*. While its attempt to change the practice of science and US public school curricula did not succeed, it nevertheless had a significant impact on popular views of science. The big tent was ultimately more of a political alliance than a theological one (even if these categories often blended), which means that those who were under its auspices were often only tenuously in agreement with each other—more united by what they opposed than by what they supported (Darwin being the most prominent foe). But increasingly, there was unified opposition regarding the role of experts in politics and government. ID may not have changed the way education or science was done in the US, but it did dramatically change the way science and scientists were seen and trusted. The big tent very much continued to exist, but the glue holding it together became suspicion and cynicism—a pervasive skepticism that ranged beyond just of the relatively narrow issue of Darwin.

5

Aftermath

What Happened after the Court Case?

"ID is dead. As a doornail." This verdict came from mathematician Jason Rosenhouse in 2011.[1] Rosenhouse was not alone in his diagnosis. Kevin Padian, a paleontologist and expert witness at *Dover*, wrote in 2019, "The Dover trial annihilated the whole operation. You have heard almost nothing from the Discovery Institute since 2005. And they were once so formidable they had op-eds in the *Wall Street Journal* and the *New York Times*."[2] The Christian physicist Paul Wallace agreed, writing, "ID has no future."[3]

Naturally, ID advocates disputed this conclusion. David Klinghoffer mocked Rosenhouse for proclaiming ID's death, writing, "Darwinists were saying the same thing back in 2005 when an obscure judge in Pennsylvania took dictation from the ACLU and declared ID to be religion instead of science, thus settling the matter forever and ever."[4] In a different response to Rosenhouse, the Discovery fellow Casey Luskin argued that ID's best years were ahead, writing, "Despite what you hear—or don't hear—from critics (especially those in the media), the past 5 to 10 years have been a boom period for pro-ID scientific research and peer-reviewed scientific publications." He lauded the founding of the BioLogic Institute in Seattle, a center for research into intelligent design as science, and contended that critics needed to "get over" *Dover*. The myopic focus on the court case was evidence of critics' desire to "convince people that there's only one side worth considering. Those who care more about scientific truth than about politics or power can rest assured that the future of ID is bright."[5]

ID did not disappear after *Dover*; its scientific projects continued, but its cultural and political campaigns became the dominant focus. ID's scientific efforts revolved around the founding of the design-centric research lab the BioLogic Institute—actually right before

Dover—and the recent establishment of the Walter Bradley Center (with its timely focus on artificial intelligence), as well as the periodic publication of work in the ID scholarly journal *BIO-Complexity*. However, while a steady stream of books and articles continued to be produced, most of this newer material, especially the public-facing material (a large portion of which was published through the Discover Institute's own press), doubled down on the arguments and evidence that had already been presented in the 1990s and early 2000s. In the years after *Dover*, Jonathan Wells, Michael Behe, Stephen Meyer, and more reiterated and reprised the same general arguments that had failed to win the day in court.

This emphasis on repeating the same arguments was partnered with a corresponding increase in cultural aggrievement and a sense of persecution on the behalf of ID's supporters. With regard to books, this trend manifested itself most clearly in Jonathan Wells's follow-up books *Zombie Science* and *The Politically Incorrect Guide to Darwinism and Intelligent Design*, but the rhetorical strategy reached its apotheosis in the ID documentary *Expelled*. The repetition of old arguments, the allegations of widespread conspiracy on the part of evolutionary scientists, and martyrdom for ID proponents became part of ID's dominant rhetorical style, especially as a popular and public strategy. In light of everything else after *Dover*, it was *Expelled* that most succinctly captured and exemplified the state of design in the twenty-first century. Over time, rhetoric can become reality. The accusations of persecution were not simply sly, political dog whistles to supporters with congruent politics. Rather, they were deeply felt and believed convictions on the part of ID advocates. The rejection of the "establishment," encapsulated in *Expelled* (and with it, the disavowal of the normal rules for scientific publishing and expertise), would not stay restricted to evolutionary theory. Instead, ID proponents began to question the prevailing scientific wisdom about vaccines, climate change, astrophysics, and AIDS. Anti-expertise became not just rhetoric but rather a hermeneutic (one of suspicion, perhaps) and the basis for a worldview deeply skeptical of scientific progress and knowledge. ID advocates had long championed the capability of science to provide concrete knowledge, but in the post-*Dover* world, this conviction wavered.

Design and the Continuing Effort at Scientific Research

Although the cultural and political wars were dominant in post-*Dover* ID, its scientific projects did continue. In 2005, before the court hearings but while the trial controversy was in the news, the BioLogic Institute was founded in Redmond, Washington, with the express purpose of pursuing scientific research from an ID perspective. Douglas Axe was appointed director of the institute. A molecular biologist with lab experience working on protein folds at the University of Cambridge, Axe was awarded a PhD from Caltech. In 2000, while at Cambridge, Axe published an article in the *Journal of Molecular Biology* that did not expressly support ID but from which he afterward drew design-friendly conclusions.[6] The ID critic and theistic evolutionist Dennis Venema described Axe's work on protein folding as his "main contribution" to ID, highlighting Axe's view on "the limits of what evolution can accomplish to produce new protein shapes (folds) and functions." Proteins need to be in certain shapes to function properly, and the number of possible functional proteins is vanishingly small compared to the total possible number of possible protein sequences. Therefore, evolution was an implausible candidate for the development of new proteins because the "search target" is too small to be stumbled on by chance.[7] Extracurricular disagreement between Axe and the Cambridge scientist Sir Alan Fersht eventually led to Axe leaving the lab in 2002. Axe later contended that Fersht "[gave] in to the internal whistle-blower who wanted [him] removed."[8] In 2005, Axe was appointed head of the new BioLogic Institute, a follow-up on the plans outlined in the Discovery Institute's controversial "Wedge Document," which in 1998 called for "front line research funding at the 'pressure points' (e.g., Paul Chien's Chengjiang Cambrian Fossil Find in paleontology, and Doug Axe's research laboratory in molecular biology)."[9]

The next year, in 2006, the science writer Celeste Biever traveled to Redmond for an investigative piece on BioLogic for *The New Scientist*. Titled "The God Lab," the article documented Biever's attempt to learn the purpose of BioLogic, but her mission was complicated by the fact that only George Weber, one of BioLogic's founding directors and a professor of business and administration at Whitworth College in Spokane, was willing to speak openly to her. Weber told Biever, "We are the first ones doing what we might call lab science in intelligent design. . . . The objective is to challenge the scientific community on naturalism." Before the article could be published, however, Axe relieved Weber of his post and explained to Biever that the former board member had "seriously misunderstood the purpose of BioLogic and [had] misrepresented it." Axe and the rest of the BioLogic Institute maintained that they had structural independence from the Discovery Institute, though they did admit to financial support from it. Biever surmised that the BioLogic Institute was a new step on ID's evolutionary path, noting that after Judge Jones had blasted ID for not being "the subject of testing and research," a lab doing just that would be a "new stage" in which skeptical scientists could search for evidence of a creator. It would also provide a much-needed publicity win after the *Dover* trial. She included a quotation from Ronald Numbers in which the historian explained, "It will be good for the troops if leaders in the ID movement can claim: 'We're not just talking theory. We have labs, we have real scientists working on this.'"[10] Other critics were far less sanguine. *Reason*'s science writer Ronald Bailey compared "Big Intelligent Design" to "Big Tobacco" and highlighted what he perceived as a similar, cynical strategy: larding less-than-prestigious journals with ideologically driven science in order to showcase doubt to the public.[11]

Biever's portrait of the "God Lab" was not flattering, but neither was it overtly hostile. Axe, Ann Gauger, and Brendan Dixon (two other researchers at the BioLogic Institute) responded with a letter to *The New Scientist*, writing, "your editorial asks a reasonable question: can the theory of intelligent design (ID) lead to good science?" The perception that the institute existed only to question Darwinism was "incomplete"—there was, for Axe and company, a positive potential in studying biology from a design vantage, and Darwinism "prevents what we have learned as engineers from illuminating biology." If design was right and accident was wrong, then "whole new fields open up, waiting to be explored." They concluded, "We don't know where all this would lead, but we are confident that good science will come out of it."[12]

As the years passed, however, the institute had its ups and downs. In 2010, BioLogic supported the launching of a peer-reviewed, ID-focused scholarly journal: *BIO-Complexity*. This would not be the first attempt at an ID-themed journal, for it had predecessors in *Origins & Design*, which ceased publication in 1999, and *Progress in Complexity, Information, and Design*, which ceased in 2005.[13] According to *BIO-Complexity*'s self-description, it aimed "to be the leading forum for testing the scientific merit of the claim that intelligent design (ID) is a credible explanation for life."[14] For the Discovery fellow Jay W. Richards, the journal was meant to resolve the "Catch-22" in which ID supporters found themselves: publicly impugned for not publishing peer-reviewed work and then denied the privilege of submitting design-friendly articles to established journals. Richards opined, "anyone with the slightest acquaintance with the subject knows that arguing explicitly for design in an article submitted to a scientific journal is a sure-fire way to prevent the article from seeing the light of day."[15] Denied entry to the club, design theorists would create their own, and the BioLogic Institute would be the primary support for the operation of *BIO-Complexity*.[16]

As with the BioLogic Institute, *BIO-Complexity* saw its share of criticism. The computer scientist Jeffrey Shallit highlighted the slow pace of publication, noting that the 2012 issue had only four contributions (with only two being research articles) while also drawing attention to the fact that three of them were written by people running the journal. He concluded that this was further evidence that ID was pseudoscience, for "pseudoscience is *sterile*: the ideas, such as they are, lead to no new insights, suggest no experiments, and are espoused by single crackpots or a small community of like-minded ideologues."[17] Shallit returned in 2014 with the same criticism, noting that this time, of the four newly published pieces, only one was a research article, and asked, "What happened to the claim that ID creationists stand for ideas? One research article a year is not that impressive. . . . Why can't their own flagship journal manage to publish any of them?"[18] Glenn Branch, the deputy director at the National Center for Science Education opined, "'Intelligent design' journals thus seem to be a scientific cul-de-sac." Furthermore, "Scientists with anything scientifically important to say about 'intelligent design' will . . . take it to the mainstream scientific literature, which is already widely disseminated and respected, not to a parvenu like *BIO-Complexity*."[19] *BIO-Complexity* still lingered, however, with a new volume published annually (ranging from two to four articles each time).

Despite the withering criticism, ID advocates continued to toil. BioLogic and *BIO-Complexity* are two of the more noteworthy attempts at pursuing science from a design standpoint. However, *BIO-Complexity*'s existence (and the existence of its doomed forebears) suggests an important development in design strategy. Any attempt to gain acceptance through the traditional mechanisms of scientific progress were quashed; the establishment of peer-reviewed design journals instead signaled the creation of an alternate ecosystem, a parallel scientific world in which design was the reigning hypothesis and where Darwinism had been displaced. Such a world existed conterminously but incommensurably with the dominant paradigm in mainstream science. In the world of ID, the arguments worked—the paradigm shifted. ID, then, did not need to change to be successful; it needed to reiterate and sustain what was already known. And so, it is not especially surprising that the most notable design arguments made in the post-*Dover* era, those presented by Stephen C. Meyer, were a replication and refining of design's first warning shot that was fired all the way back in 1984 with *The Mystery of Life's Origin*.

Stephen Meyer and the God Hypothesis: Intelligent Design's First and Last Argument

In a discussion of Douglas Axe's work on protein folds, the philosopher Vincent Torley wrote that Axe encapsulated the case for ID in a single sentence: "functional coherence makes accidental invention fantastically improbable and hence physically impossible." This compact statement was at the heart of nearly all intelligent design arguments, from Behe to Dembski to Axe. Axe's specific work on the long odds of building a 150-amino-acid protein by chance was crucial. Continued Torley, "The importance of this particular argument to the case for Intelligent Design cannot be over-emphasized. Putting it succinctly: if it fails, then we're back at square one, in terms of building a mathematical case for ID." In fact, this argument would be central to two of the most noteworthy ID books produced in the post-*Dover* era, both by Stephen Meyer. As Torley wrote, "Dr. Stephen Meyer's two Intelligent Design best-sellers, *Signature in the Cell* and *Darwin's Doubt*, are built on the bedrock foundation of this argument: their whole case would collapse without it."[20]

Stephen Meyer was there at the genesis of ID. In his own account, he was a geophysicist working for a multinational oil company until 1985, when by chance he attended a conference in Dallas where he would become involved with the ID movement at the ground floor. He observed a panel discussion on *The Mystery of Life's Origin* and then managed to meet with Charles Thaxton, one of its authors. He was fascinated by the controversy around the book and particularly its epilogue, which briefly suggested the possibility that a divine designer could be a scientific hypothesis. Meyer listened to a panel debate between *Mystery*'s authors, as well as Dean Kenyon, and the well-known evolutionary biologist Russell Doolittle on whether abiogenesis was feasible. "I left deeply intrigued," wrote Meyer, and he was taken with the "radical claim that an intelligent cause could be considered a legitimate *scientific* hypothesis for the origin of life." The event changed all of Meyer's professional priorities. "By the end of that year," he wrote, "I was preparing to move to the University of Cambridge in England, in part to investigate questions I first encountered on that day in February." Meyer would go on to earn a doctorate in the history and philosophy of science at Cambridge.[21]

In the years that followed, Meyer would be a persistent presence in the ID world. He was part of the original Ad Hoc Origins Committee that defended Johnson's *Darwin on Trial* after publication, and he also authored an article in *The Wall Street Journal* in 1993 that defended Dean Kenyon.[22] He helped found the Center for Renewal of Science and Culture (now without the "Renewal") in 1996.[23] He testified on behalf of intelligent design at the Kansas evolution hearings in 2005.[24] Beyond all this, though, Meyer's name truly rose to the forefront of ID when in 2004 he managed to publish the first peer-reviewed intelligent design article in a scholarly journal. The ensuing fracas led to an investigation into the peer-review process of the journal, the *Proceedings of the Biological Society of Washington*, and the eventual censure of its editor, Richard Sternberg. Following the "Sternberg affair," as well as the fallout of the *Dover* trial, Meyer decided to wade into the popular book front and write his own treatise on intelligent design. He sought to correct the record—he wanted to clarify ID's differences from creationism and clear up misconceptions about design-as-such (for instance, that it did not necessarily deny common descent and could even plausibly countenance a kind of "front-loaded," directed evolution; rather, it denied purposelessness). Furthermore, he hoped to introduce the public to a separate argument for ID, one that was related to but not based on Behe's or Dembski's work but that instead focused on the old mainstay of chemical evolution and the alleged impossibility of abiogenesis—either by chance or by self-organization. Meyer related that he chose to write a popular book rather than an academic article because a trade book would "go over the heads of an entrenched establishment to force a reevaluation of an established theory by creating wider interest in its standing."[25] In 2009, that book would be released under the title *Signature in the Cell*.

"The problem of the origin of life is clearly basically equivalent to the problem of the origin of biological information," Meyer quoted Bernd-Olaf Küppers.[26] Resolving this question in favor of intelligent design would be the central thrust of *Signature in the Cell*. Meyer defended his role as an outsider, one who was not a practicing scientist, by pointing to Crick and Watson, the discoverers of the structure of DNA. They were fresh blood, achieving their enormous feat "not by working their way up through the establishment" and engaging in the rat race of paper publishing but instead "explaining an array of preexisting evidence in a new and more coherent way."[27] Meyer recapped the work of Claude Shannon on information theory, then turned to William Dembski and Douglas Axe, before running through the various hypotheses for chemical evolution—the "classic" Oparin model, DNA-first theories, and the more recent RNA World model, as well as "self-organization" theories advocated by scientists such as Stuart Kauffman—and finding all of them wanting.[28] Meyer made much of design as an alternative hypothesis based on abductive reasoning—the "inference to the best explanation"—as conceived by the philosopher Charles Sanders Peirce (1839–1914). Furthermore, for Meyer, design fell within properly constructed "historical science" (which was different from "experimental science"), a form of investigation that, as he borrowed from Stephen Jay Gould, was "testable, but not necessarily by experiments under controlled laboratory conditions." Rather, historical science meant testing theories by "evaluating their explanatory power."[29] For Meyer, design was precisely the hypothesis that boasted the best explanatory power of the facts of the distant past. In an inversion of classic young-Earth creationist rhetoric, Meyer took great pains to establish the importance of uniformitarianism as a scientific principle, arguing that design advocates were as thorough in their adherence to it as were Charles Lyell and Charles Darwin.[30] Because design was a common experience in day-to-day life in the present, and—as Meyer argued—it was the only known source of "information-rich systems," it was therefore legitimate to project that knowledge onto the prehistoric world and conclude that design was the "best explanation."[31] The present, as the uniformitarian saying goes, was the key to the past. This reliance on uniformitarianism also, for Meyer, set ID far apart from young-Earth creationism, which virtually unanimously rejected the validity of the principle. It further distinguishes Meyer from proto-ID works like Geisler's *Origin Science*, which still adhered to an older creationist belief in the untestability of past "singularities."

Signature in the Cell received both surprising praise and familiar criticism. The eminent philosopher, and outspoken atheist, Thomas Nagel nominated the book to *The Times Literary Supplement* as a "Book of the Year" for 2009, writing that "Meyer is a Christian, but atheists, and theists who believe God never intervenes in the natural world, will be instructed by his careful presentation of this fiendishly difficult problem."[32] Nagel's action triggered an immediate but opposite reaction, with the chemist Stephen Fletcher replying in a letter to the editor that intelligent design's central supposition of supernatural action was equivalent to primitive belief in spirits monkeying with everyday affairs and that "for a society with no concept of bacteria, this is, perhaps, a forgivable conceit. But for a modern university professor to take this idea seriously is, I think, mind-blowing." Nagel himself surmised that Fletcher had not read *Signature* (for "it includes a chapter on 'The RNA World' which describes that hypothesis for the origin of DNA at least as fully as the Wikipedia article that Fletcher recommends") and argued that "the tone of Fletcher's letter exemplifies the widespread intolerance of any challenge to the dogma that everything in the world must be ultimately explainable by chemistry and physics."[33] A few years later, Nagel would pen his own anti-Darwinist book, *Mind and Cosmos*, in which he argued that reductive materialism could not explain existing questions in philosophy of mind and could not explain consciousness. He defended ID, writing, "Even if one is not drawn to the alternative of an explanation by the actions of a designer, the problems that these iconoclasts pose for the orthodox scientific consensus should be taken seriously. They do not deserve the scorn with which they are commonly met. It is manifestly unfair."[34] The split between the insider and outsider was evident in the response, for while *Signature* was popular with critics of

Darwin, it received little notice by the academic community. At *Panda's Thumb*, Richard Hoppe noted that none of the reviews had appeared in scientific journals or popular science magazines, perhaps highlighting the paradox that *Signature* made public noise but academic critics met it with "crickets," in Rosenhouse's words.[35]

Meyer responded to his interlocutors, and David Klinghoffer compiled a book regarding the controversies; but the fullest defense would come a few years later, in 2013, when Meyer released another book, expanding his attack on Darwinism from chemical evolution to evolution as a whole.[36] From Meyer's perspective, *Signature* had manifestly not been an attack on evolution but rather focused on the question of abiogenesis. Wearied by critics accusing him of anti-evolutionism, Meyer wrote, "I found this all a bit surreal, as if I had wandered into a lost chapter from a Kafka novel." Never the worse for wear, however, Meyer returned to do what critics alleged he had already done. "The repeated prodding of my critics has paid off," he wrote; "even though I did not write the book or make the argument that many of my critics critiqued in responding to *Signature in the Cell*, I have decided to write that book."[37] That book would be *Darwin's Doubt*. Much of Meyer's argument was a recapitulation of *Signature*. Though he focused on development of animal body plans and the problem of transitional fossils in the Cambrian era (or lack thereof), he also trod the same paths as before, relying on Douglas Axe's work on protein folds and on Michael Behe and David Snoke's 2004 paper regarding proteins, genes, and irreducible complexity, as well as Axe and Ann Gauger's work at BioLogic.[38] At several points, Meyer directed the reader to refer back to *Signature* for a fuller explanation of the argument.[39] *Darwin's Doubt* was derived largely from *Signature*, but *Signature* itself was a rejuvenation of the arguments made by Thaxton and company in *The Mystery of Life's Origin*. The old controversies were being relitigated in the post-*Dover* era, and Meyer would not be the only one to give them a fresh coat of paint.

Doubling Down: The Arguments of Yesterday Made Again Today

The repeated reuse of arguments from *Signature in the Cell*, dusted off and repeated again in *Darwin's Doubt*, was not an isolated incident but became part of many intelligent design theorists' dominant argumentative style in the late aughts and teens of the twenty-first century. Even *Signature*, arguably the most important ID book after *Dover*, was described by its own supporters as a natural follow-up to the arguments first made in 1984's *The Mystery of Life's Origin*. Indeed, Klinghoffer called *Mystery* a "daring first draft" of what would eventually become *Signature in the Cell*.[40] *Signature* was not alone in this: nearly all of the material produced by ID's leading lights after 2005 were recapitulations, replications, and restatements of prior arguments—notably Michael Behe's *Edge of Evolution* and *Darwin Devolves*, Jonathan Wells's *Zombie Science*, and, bringing ID all the way back to the beginning, the 2020 reissue of *The Mystery of Life's Origin*. All of these books were printed by popular and not academic presses, as it had been the ID mantra since Johnson (taken up by Behe, Meyer, and Wells, most prominently) to take the argument out of the lab and appeal directly to the masses in a populist flanking maneuver against the scientific establishment.

Despite the public failure in court, Behe was not about to give up on irreducible complexity. It reappeared in his 2007 book *The Edge of Evolution*, in a slightly modified form. Though critics accused him of "throwing up [his] hands" at the problem of the flagellum motor's evolution, Behe asserted, "They may say it again . . . but the discoveries of the past decade have made the problem worse, not better, both at the level of protein machinery and at the level of DNA instructions."[41] Despite the promises of Richard Dawkins and Kenneth Miller, Behe wrote that there was still no plausible pathway to ascertain the evolution of the flagellum or cilia, merely some just-so stories. "In the more than ten years since I pointed it out," he wrote, "the situation concerning missing Darwinian explanations for the evolution of the cilium is utterly unchanged."[42] Unlike in *Darwin's Black Box*, however, Behe made more space in *Edge* for a front-loaded view of directed evolution (or "extended fine-tuning"), one that was guided by a continuously intervening God (which he distanced from theistic evolutionists, who he argued believed only in a noninterventionist deity).[43] Richard Dawkins, in a *New York Times* review of *Edge*, repeated his lofty disdain for ID and contrasted *Edge* with *Black Box*: the first book was "enlivened by a spark of conviction, however misguided," whereas Behe's second foray was "the book of a man who has given up. Trapped along the path of his own rather unintelligent design, Behe has left himself no escape."[44]

Behe kept the faith with irreducible complexity, however, and it made an appearance again in his 2019 return to design, *Darwin Devolves*. All the familiar faces—the mousetrap analogy, the bacterial flagellum, the eye—gathered for a reunion, and Behe contended that "the difficulties [irreducible complexity] presents for Darwin's theory have grown much worse in the past several decades." Furthermore, despite ample time for them to do so, "opponents of intelligent design have tried and failed to find a plausible Darwinian route even to a simple mechanical mousetrap." Behe did modify the argument a bit, including a newer angle on "mini" irreducible complexity (complexity on a smaller scale), such as a hook-and-eye latch or when two cysteine groups form a disulfide bond.[45] Behe did not only reprise *Black Box*, either, for *Darwin Devolves* also expanded on *Edge*'s argument that evolution had limits, and indeed was "self-limiting," and then posited that natural selection and random mutation caused organisms to degrade and devolve rather than improve and evolve.[46] Jerry Coyne recognized that Behe's views were not strictly creationist, that he did not base them on biblical interpretation, and that he was something of a teleological evolutionist but criticized the book for nitpicking Darwin and failing to supply a robust alternative, writing, "Well, now it's 20 years on, and despite the efforts of Behe and other neo-creationists, intelligent design has been discredited as science and outed as disguised religion." He also noted the significance of the book's publication by HarperOne, "the religious, spiritual, and self-help division of HarperCollins."[47] The biologist Nathan Lents responded in *Skeptic*, in a cover article titled "Behe's Last Stand," complete with a dramatic image of a rather bemused Charles Darwin holding Behe in his hand, while the tiny Behe lectures at the legendary scientist. Lents argued that Behe "continues to dig himself further into the hole he opened 20 years ago with *Darwin's Black Box*." Plenty of scientists addressed the biochemical challenge from that book, wrote Lents, and Behe "may find the hundreds of published refutations unconvincing, but the onus is on him to explain why. Instead, he pretends they don't exist."[48]

The trend toward doubling down was most explicit in another ID luminary's repetition of previous arguments: Jonathan Wells and his icons. Wells achieved prominence at the millennium with a bold attack on evolutionary theory that revolved around the presentation of familiar proofs in textbooks. *Icons of Evolution*, as the 2000 book was called,

Figure 5.1. *Skeptic* magazine cover in 2019. (Lents, "Behe's Last Stand")

not only alleged widespread misinformation but also argued that many of the famous "icons" of evolution—like Haeckel's embryos or the peppered moth experiment—were known to be inadequate by the writers of the textbooks themselves. Wells wrote that Jerry Coyne learned of the flaws in the peppered moth experiment much later in his career, and the accomplished biologist compared it to discovering "that it was my father and not Santa who brought the presents on Christmas Eve."[49] For Wells, the ubiquitous presence of these icons in textbooks indicated a conspiracy on evolutionary biology's part—whether maliciously intentional or simply incidental. The bigger issue for him was that "the general public is rarely informed of the deep-seated uncertainty about human origins that is reflected in these statements by scientific experts." Wells concluded by scandalously reporting that dead-end evolutionary research was funded by government agencies like the National Institutes of Health and NASA and so "they're doing it with your money."[50] Coyne took exception and noted in *Nature*, "Authors of some biology texts may occasionally be sloppy, or slow to incorporate new research, but they are not duplicitous." He drew attention to the irony of Wells's intimation that there was a campaign of suppression when "it is invariably evolutionists (including [Coyne himself]) who have noted problems with some classic icons of evolution."[51]

Wells wrote a sequel to *Icons* in 2017 called *Zombie Science*, which was subtitled *More Icons of Evolution* and bore the familiar image of "the descent of man" but ending in a pale, undead shuffler holding a sketch of the tree of life, along with the slogan "they just keep coming back!" Coming on the heels of the total saturation of pop culture with

Figure 5.2. Front cover of Jonathan Wells's *Icons of Evolution: Science or Myth? Why Much of What We Teach About Evolution Is Wrong* (2000).

Figure 5.3. Front cover Jonathan Wells's *Zombie Science: More Icons of Evolution* (2017).

zombie-themed postapocalypticism (stretching from *The Last of Us* to *The Walking Dead*), "zombie science" was for Wells the persistent presence in science that was actually materialist philosophy.

Wells documented a list of the same supposedly flawed icons as before, including the familiar "tree of life" regarding common ancestry, vestigial organs like the appendix, the evolution of the eye, the origin of DNA, and the transitional species between land animals and whales. He expanded to some new frontiers as well, ending on a short discussion of medicine and its debt (or, for Wells, lack thereof) to Darwin (though this was presaged in his earlier book *The Politically Incorrect Guide to Darwinism and Intelligent Design*). Wells responded to Coyne's review and contended that Coyne knowingly used disproved icons, like Haeckel's embryos and peppered moths, in his popular book *Why Evolution Is True*.[52] *Zombie* concluded with a much more dismissive and conspiratorial tone than the original *Icons*, as the older book made space for unthinking conformity—rather than outright fraud—as a plausible reason for the endurance of the icons. Instead, in *Zombie*, Wells attributed ID's failures to the innate hostility of "establishment science" and cast aspersion on the honesty and capability of contemporary scientists, writing that an increase of retractions in journals like *Nature* could be the result of "hasty or careless work, but a growing percentage of retracted papers result from deliberate fraud."[53] "Materialistic science . . . is corrupting empirical science," concluded Wells, but intelligent design could salvage science. Indeed, for Wells, ID was growing "despite fierce opposition from establishment science."[54]

The apex of design redoubling may be the 2020 reissue of *The Mystery of Life's Origin*, reprinted and expanded three and a half decades after its original appearance in 1984. The new edition featured an in-depth introduction written by David Klinghoffer that included the story of *Mystery*'s road to publication and how it eventually came to find its publisher as well as the support of Dean Kenyon. *Mystery*, for Klinghoffer, contained all that was new in ID and would later get expanded, as well as all that distinguished it from scientific creationism (particularly its affirmation of an old Earth and uniformitarianism). *Mystery* adumbrated all the arguments that would be seen later: the principle of uniformity, applying what "we know by experience" about the behavior of "intelligent investigators," Shannon information, and "specified complexity."[55] The main text of the reissue remained almost unchanged from its original form. "The fact that only light updating was needed," wrote Robert J. Marks and John G. West in the foreword, "is a testament to the meticulous scholarship of the authors and to the enduring nature of the problem they identified in origin of life studies."[56] After the main text, however, was a 1997 update written by Charles Thaxton, which first appeared in the Hungarian edition. Thaxton noted that nothing had changed their position, though he lamented that the media often presented origin of life studies as a thriving science rather than a terminally ill one. Newer hypotheses, like the RNA World, were rejected, and Thaxton proposed to "reexamine" the design "hypothesis." The difficulty was in getting a hearing for ID, as the idea was usually opposed due to kneejerk principle. Once again, however, in a point that would be taken up by Meyer in *Signature*, Thaxton rejected the distinction between operations and origins science and affirmed uniformitarianism. Design was a principle at work today; the past was no different.[57]

The second section of the book, "The State of the Debate," included short essays regarding the controversy between design and Darwin, with James M. Tour providing an overview of the successes (or, more commonly, the failures) of origin of life research, Guillermo Gonzalez writing a perspective on what astrobiology—that is, the study of life and how it may have arisen on other planets—could teach about the origin of life on Earth, and an essay on thermodynamics and abiogenesis from Brian Miller.[58] Jonathan Wells reprised another exposé of the icons of evolution, with the Miller-Urey experiment taking center stage due to its prevalence in textbooks as an example of the generation of life from nonlife. Wells noted that despite the flaws in the experiment (and Miller's follow-up in 1983), it was still a widely disseminated bit of textbook orthodoxy and was defended as a definitive demonstration of abiogenesis.[59] The reason for its continued presence was, for Wells, the same "zombie science" he wrote about in his own book—that is, "a dogmatic commitment to materialistic philosophy. Biology courses were being misused to indoctrinate students in materialism."[60] The book concluded with a chapter from Stephen Meyer, a précis of *Signature in the Cell* that ran through the complexity of the cell, the importance of information, the implausibility of chance theories and also of self-organization, and the function of design as the inference to the best explanation. In the self-organization section, Meyer directed the reader to *Signature in the Cell* for further reading.[61]

The repeated reduplication of familiar material and argumentation brought with it a far more conspiratorial tone and air of grievance. In Tour's chapter, for instance, the media was singled out for toeing the party line regarding Darwinism and for being overzealous in trumpeting any new discoveries in origin of life studies. Such excitement duped the average person into thinking that it was a thriving field, unlocking life's mysteries, when for Tour it was not. He included what he took to be the "how-to" guide for those who were undertaking chemical synthesis experiments: after performing the experiment, "publish a paper making bold extrapolations about [the] origin of life," "engage with the often over-zealous press to dial up the knob of unjustified origin-of-life projections," and "watch the layperson being misled."[62] Tour was not alone in this sense of slight from the media. In the post-*Dover* era, an explanation for ID's failure in persuading both the scientific establishment and wider world was needed. An answer was ready and clear: persecution.

The Persecution of Design

Michael Shermer, editor in chief of *Skeptic* magazine and avid bicycle enthusiast, sits on the movie screen across the table from Ben Stein, the conservative speechwriter, commentator, and actor. Stein is quizzing Shermer on the fairness of design's treatment in public and academic circles. The topic is the alleged mistreatment of Richard Sternberg, the editor of the *Proceedings of the Biological Society of Washington*, who drew the ire of the scientific community for publishing the aforementioned Stephen Meyer article in 2004, the first such publication for the intelligent design community. Shermer responds, "People don't get fired over something like that. You roll up your sleeves, you get to work, you do the research, you get your grants, you get your data, you publish, and you work your butt off, and that's how you get your theories taught." Stein, unpersuaded, asks, "What if you try and try and roll up your sleeves and go to work, and work your butt off, and they say, 'we're going to fire you if you even mention the word 'intelligent design'?" Shermer shakes his head: "I don't think that's happened. Where is that happening?"[63]

This scene played out in the 2008 intelligent design documentary *Expelled: No Intelligence Allowed.* Cowritten by Stein and directed by Nathan Frankowski, *Expelled* was a pro-design roundup of various examples of persecution. The substance of the film focused more on politics and social mores than it did on science, with very little intelligent design theory making its way into the movie. Rather, attention was paid to the uniform hostility of the scientific establishment and the moral corrosiveness of Darwinism. Stein continues a vaunted trend of characterizing the anti-evolutionary struggle as essentially democratic, a small minority pleading for their voice to be heard and not drowned out by the tyranny of the majority. "Freedom is what makes this country great," Stein addresses a crowd at the film's start: "Freedom has allowed us to create, to explore, to overcome every challenge we have faced as a nation. But imagine if these freedoms were taken away. Where would we be? What would we lose? Well, unfortunately, I no longer need to imagine. It's happening. We are losing our freedom in one of the most important sectors of society: science."[64] As we have seen before, ID supporters found in Thomas Kuhn a framework and historical blueprint with which to understand themselves, including an explanation for why the devotees of the reigning Darwinian paradigm would react with, in Kuhn's word, "intolerance" to dissenters. *Expelled* focused entirely on this process. Bruce Chapman, in describing the movie, contended that ID was targeted for destruction. "Most Darwinists have avoided debates," he wrote, "and in universities have stooped to denial of academic tenure, promotions and even graduate-student status to dissenters."[65] These are the titular "expelled." The film chronicles a few victims of implacable Darwinian fury: Richard Sternberg, Caroline Crocker, Guillermo Gonzalez, Michael Egnor, and Robert J. Marks II. Conspiratorial narration saturates the movie. It is constructed around a narrative framework that highlights the various persecutors of ID—the media, the courts, the education system—and how each one helped prevent a proper hearing.[66]

One of the earliest such instances of professional persecution—one that, although it did not make it into *Expelled*, set the tone for much of ID's subsequent relationship to the scientific community—was the October 1999 establishment, and subsequent dissolution, of William Dembski's Michael Polanyi Center at Baylor University. Under the auspices of Baylor's new president, Robert B. Sloan, controversy brewed when in 2000 the center held the "Nature of Nature" conference, triggering discontent from many of Baylor's science faculty. The neuroscientist Lewis Barker stated that there was "unanimous consent that the Polanyi Center is detrimental to Baylor's science department."[67] The conference was held, and many scientists—including those skeptical of ID—attended, but the center would not last long afterward. Baylor's faculty senate voted 27–2 to recommend the research center be dissolved, but President Sloan rejected the recommendation. In his 2000 "State of the University" speech, Sloan (who had previously quashed creationist attempts to get evolution barred from Baylor) argued that the problem critics had with the center was not the process through which it was established (some critics alleged that he had not followed procedure) but was rather an interest in censorship. This issue was one of "intellectual and academic integrity." Baylor would not balk at questions that were "politically incorrect," and Sloan was "proud of Baylor's willingness to ask questions which some are apparently afraid to entertain."[68]

A follow-up External Review Committee, which convened in September 2000 to advise the administration on what to do, concluded with a compromise: Dembski's work could continue under the more interdisciplinary Institute for Faith and Learning, but the committee also firmly recommended that Polanyi's name be removed from any group or center that focused on intelligent design, since the Hungarian philosopher had rejected a similar concept in *Personal Knowledge*.[69] Dembski felt vindicated and so issued a widely disseminated email, declaring, "this is a great day for academic freedom" and "dogmatic opponents of design who demanded the Center be shut down have met their Waterloo. Baylor University is to be commended for remaining strong in the face of intolerant assaults on freedom of thought and expression."[70] Dembski was subsequently removed from his position. Baylor officials explained in a press release that the review committee report had called for a compromise and highlighted the need for "collegial" behavior. The Institute for Faith and Learning's director, Michael Beaty, stated that "Dr. Dembski's actions after the release of the report compromised his ability to serve as director."[71]

Though Dembski's response was his undoing, his fate was highlighted by design advocates for years as an example of unjust persecution. In a piece in *The American Spectator*, Dembski's termination was described as a "lynching," and the author related Dembski's belief that "the siren call of the internet" had stoked the flames of scientific retaliation.[72] Jonathan Wells commented on the events in both *Icons of Evolution* and *The Politically Incorrect Guide to Darwinism and Intelligent Design*, approvingly noting that Congressman Mark Souder had censured the Baylor scientists and that Sloan had stated that their actions "border[ed] on McCarthyism." "Darwinists will not tolerate any open discussion of intelligent design," wrote Wells; "when they cannot crush it in a balanced academic forum, they resort to intimidation and mob rule."[73] In the Discovery Institute's response to Judge Jones's *Dover* ruling, Dembski was included in a list of wrongly persecuted design advocates. That he was "banned from teaching" at Baylor was blamed on Barbara Forrest and her attempts "to dissuade scholars from associating with Dembski's Polanyi Center."[74]

In these accounts, sympathizers downplayed Dembski's own role in his termination. However, even some of those who were inclined to sympathy for design rebuked him. As Francis Beckwith saw it, "Instead of offering an olive branch and conciliatory tone at the moment of victory, Dembski angered many faculty members and embarrassed his benefactors and supporters at Baylor. . . . In short, Dembski snatched defeat out of the jaws of victory."[75] Thomas Woodward noted that Dembski "seemed to gloat" after the committee report and that his "confrontational language" was "too much additional aggravation to an already strained network of relationships."[76] Such mixed feelings might explain why the Polanyi Center affair was not included in *Expelled*.

The subsequent brouhaha over Richard Sternberg and the publication of Stephen Meyer's pro-ID article in a peer-reviewed journal in 2004 was arguably one of the most iconic standoffs between design and evolution (second only to court hearings). *Expelled* made Sternberg the focal point of the movie's thesis of systemic persecution. "ID has failed to gain acceptance in the scientific community," wrote Judge Jones in his *Dover* decision; "it has not generated peer-reviewed publications, nor has it been the subject of testing and research."[77] While Axe's BioLogic Institute was in some respects the answer to the charge of lack of research, ID advocates long complained of bias and injustice when it came to the publication front. Wells called it a "Catch-23," writing, "Intelligent design is not scientific, so it can't be published in peer-reviewed scientific journals. How do we know it's not scientific? Because it isn't published in peer-reviewed scientific journals."[78] But, as we have seen, there was one lone exception: Stephen Meyer's 2004 article in the *Proceedings of the Biological Society of Washington*, of which Sternberg was the editor at the time. "The Origin of Biological Information and the Higher Taxonomic Changes" was a literature review regarding the Cambrian explosion, featuring many of the same criticisms of evolution that Meyer would make in later books, and concluded with a mention of design: "An experience-based analysis of the causal powers of various explanatory hypotheses suggests purposive or intelligent design as a causally adequate—and perhaps the most causally adequate—explanation for the origin of the complex specified information required to build the Cambrian animals and the novel forms they represent."[79] The reaction was swift. The next month, the council of the society released a statement declaring, "Contrary to typical editorial practices, the paper was published without review by any associate editor; Sternberg handled the entire review process." Furthermore, the society agreed with the American Academy for the Advancement of Science that "there is no credible scientific evidence supporting ID as a testable hypothesis to explain the origin of organic diversity."[80] Sternberg defended himself, writing, "I did not act unilaterally or surreptitiously in my handling of the Meyer paper," and adding that he "followed the standard review process."[81] He responded by filing a religious discrimination lawsuit against the Smithsonian, where he was an unpaid research associate. A congressional investigation headed by Mark Souder (reappearing after the Polanyi crisis) alleged that the officials in the Smithsonian Institution's National Museum of Natural History "conspired with a special interest group on government time and using government emails to publicly smear Dr. Sternberg" and that "the hostility toward Dr. Sternberg at the NMNH was reinforced by anti-religious and political motivations."[82] David Klinghoffer wrote in *The Wall Street Journal* that Sternberg was sought out like a heretic and had had his career "all but ruined" as a result.[83] In *Darwin's Doubt*, Meyer claimed that Sternberg was targeted for a campaign of harassment because he violated the scientific dictum of methodological naturalism, and Meyer drew attention to James Shapiro's earlier praise of Sternberg's work as proof that he had done quality scientific work in the past.[84]

In *Expelled*, Sternberg describes his fate at length, telling Ben Stein, as they both sit across from the Smithsonian building where his office had been, "It was all couched in terms of religion, politics, and sociology. The way the chair of the department put it is that I was viewed as an intellectual terrorist." Later, Stein bemusedly responds, "You were a bad boy. . . . You questioned the powers that be." Stein narrates, "What was Dr. Sternberg's crime? He dared to publish an article by Dr. Stephen Meyer. . . . The paper ignited a firestorm of controversy, merely because it suggested intelligent design might be able to explain how life began. As a result, Dr. Sternberg lost his office, his political and religious beliefs were investigated, and he was pressured to resign. The questioning of Darwinism was a bridge too far for many."[85] Along with the persecuted figures, such as Sternberg, the stories of personal martyrdom were interspersed with extensive totalitarian imagery. The opening credits sequence for *Expelled* features grim black-and-white imagery of walled Berlin with a score of stringed instruments overlaying them. World War II also comes into frequent focus. During the film, Stein makes repeated connections between social Darwinism, evolution, and Nazism. This association is reiterated when, in conversation with David Berlinski, the mathematician informs Stein that while Darwinism was not a sufficient condition for Nazism, it was a "necessary condition."[86] In the film's chronology of science and society, both communism and Nazism are ultimately bound up with Darwin. Critics of the film were irritated at what they perceived as the inflation of grievance and intolerance, with Ronald Bailey, for instance, writing in *Reason* that Sternberg "still has office space in the museum and has been reappointed for three more years. True, some of his colleagues might not want to hang out with him anymore. But that is a far cry from the grim black-and-white shots of Soviet armies and concentration camps featured in the film."[87]

The other figures in the film, Guillermo Gonzalez, Caroline Crocker, and Michael Egnor, recounted similar conflicts. *Expelled* highlighted Gonzalez, an astronomer who was denied tenure at Iowa State, to show the professional consequences of supporting design, as the movie alleged that Gonzalez's design publications ensured the denial of his tenure. Gonzalez's sin was that he authored the book *The Privileged Planet* in 2004, and his public defense of design earned him the enmity of another Iowa State professor, the religious studies scholar and atheist Hector Avalos, who circulated a petition at Iowa State to oppose "all attempts to represent Intelligent Design as a scientific endeavor."[88] "I worried about my tenure a little bit in 2005 when the petition was being circulated," Gonzalez tells Ben Stein, "because I viewed that as a strategy of Hector Avalos and his associates to try to poison the atmosphere on campus against me. Because he knew I wasn't tenured yet and I was very vulnerable. I have little doubt that I would have tenure now If I hadn't done any professional work on intelligent design."[89] Jonathan Wells recounted his frustration with these events in *The Politically Incorrect Guide*, writing, "So Iowa taxpayers are spending hundreds of millions of dollars every year on an institution that entrusts the teaching of religion to someone who tells students they're better off without it—and who thinks it's his job to tell an astronomer how to do science."[90] Over a decade later, and reflecting a linguistic shift in American culture, ID proponents described themselves as the first targets of cultural "cancellation." "It's safe to say," wrote Casey Luskin, "that ID proponents were being 'cancelled' by critics long before 'cancel culture' was widely known as a thing."[91] In 2021, the physicist Eric Hedin published a book called *Canceled Science*, wherein he chronicled his career trouble at Ball State University after Jerry Coyne discovered that he was teaching a class that, in part, examined ID. While presenting his side of the story, Hedin also defended ID as a whole and discussed the other instances of persecution. Sternberg, for instance, got "expelled" from the academy for his support of ID, and "this was *cancel culture* before cancel culture was a household term."[92]

Allegations of persecution do, however, make for exciting drama. *Expelled* grossed nearly $8 million in its theatrical run in 2008 and has reached half that in DVD sales since.[93] Persecution narratives are popular—Americans love an underdog—and appeals to democratic virtue, to hearing both sides, to thwarting majoritarianism likewise find wide embrace. While ID proponents often lamented the apparent hostility they received from the scientific world, they leaned into it with regularity. The back cover of the anniversary reissue for Johnson's *Darwin on Trial*, for example, boasted in large font that it was the "classic book that rocked the scientific establishment!" For Douglas Axe, the thrill of rebellion was an appealing part of working on design. "On top of the obvious intellectual importance," he wrote, "was the danger-sport-like adrenaline rush that comes from being a scientific renegade."[94] The Finnish biotechnologist Matti Leisola's recent pro-design autobiography was titled simply *Heretic* (a curious self-application for a movement known for its religious associations).[95] Not only were the leaders of ID rebels with a cause, but so were their readers. "You are about to read a dangerous book," intoned Michael Behe solemnly in the 2010 foreword to the *Darwin on Trial* reissue.[96] In the acknowledgments for *Zombie Science*, Jonathan Wells thanked "students who must remain anonymous lest enforcers of the scientific consensus destroy their very promising careers." In the introduction, there was a big, gray warning: "This book is politically incorrect, even dangerous. If you are seen reading it on a college campus, your career could suffer." There were instructions in a supplement for how to disguise the book with a paper cover, so as to avoid the watchful eyes of the campus thought police.[97] Wells was a striking figure to make such recommendations, too, for he earned his reputation for

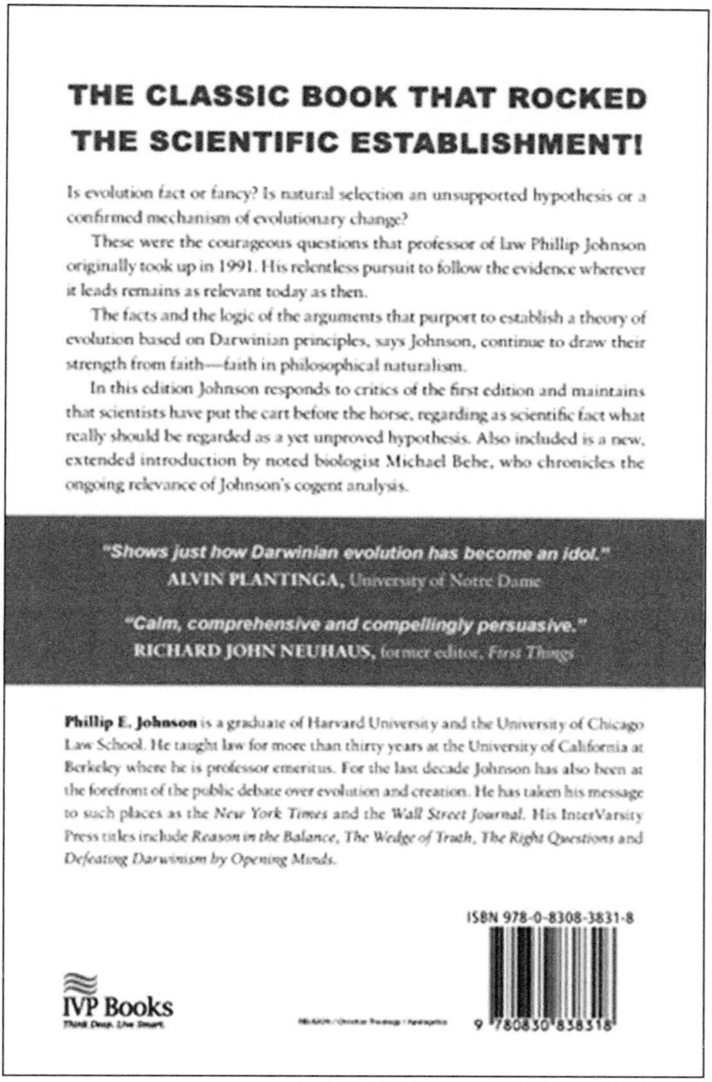

THE CLASSIC BOOK THAT ROCKED THE SCIENTIFIC ESTABLISHMENT!

Is evolution fact or fancy? Is natural selection an unsupported hypothesis or a confirmed mechanism of evolutionary change?

These were the courageous questions that professor of law Phillip Johnson originally took up in 1991. His relentless pursuit to follow the evidence wherever it leads remains as relevant today as then.

The facts and the logic of the arguments that purport to establish a theory of evolution based on Darwinian principles, says Johnson, continue to draw their strength from faith—faith in philosophical naturalism.

In this edition Johnson responds to critics of the first edition and maintains that scientists have put the cart before the horse, regarding as scientific fact what really should be regarded as a yet unproved hypothesis. Also included is a new, extended introduction by noted biologist Michael Behe, who chronicles the ongoing relevance of Johnson's cogent analysis.

"Shows just how Darwinian evolution has become an idol."
ALVIN PLANTINGA, University of Notre Dame

"Calm, comprehensive and compellingly persuasive."
RICHARD JOHN NEUHAUS, former editor, *First Things*

Phillip E. Johnson is a graduate of Harvard University and the University of Chicago Law School. He taught law for more than thirty years at the University of California at Berkeley where he is professor emeritus. For the last decade Johnson has also been at the forefront of the public debate over evolution and creation. He has taken his message to such places as the *New York Times* and the *Wall Street Journal*. His InterVarsity Press titles include *Reason in the Balance*, *The Wedge of Truth*, *The Right Questions* and *Defeating Darwinism by Opening Minds*.

ISBN 978-0-8308-3831-8

9 780830 838318

Figure 5.4. Back cover of Phillip E. Johnson's *Darwin on Trial: 20th Anniversary Edition* (2010).

rebelliousness—drafted into the army as a young man, he protested the Vietnam War and was imprisoned for it, even spending brutal months in solitary confinement.[98] All told, where design was concerned, persecution and intolerance were frequent laments, but often with it was the romantic thrill of rebellion, revolution, and sticking it to the man with one's samizdat science.

Experts and Expertise

Concurrent with the sense of persecution was a hostility toward the scientific world as an entrenched and dogmatic "establishment," an elite cadre of experts and technicians walled off from the real world of the American people. Since most Americans seemed to be skeptical of evolution, it was an act of unfreedom and illiberality that they found their views consistently thwarted by an elitist minority of academics and technocrats. A strain of anti-expertise has been present in ID's rhetoric since the beginning, especially when ID advocates courted the populist, rural, religious Americans as political allies (with sometimes disastrous consequences for ID, as the *Dover* debacle demonstrated), but it came to the forefront after 2005 due to ID's failure to gain a hearing in scientific journals, to establish and maintain consistent lab work, and to persuade the courts of its validity. All of which contributed to a doubling down on old evidence and allegations of widespread persecution. The question remained, though: Who was doing all this persecuting? Who were the guardians at the gates refusing to grant ID a hearing? The answer was invariably the establishment, the experts.

In the foreword to *Mere Creation*, the famed chemist Henry Schaefer cautioned that the "division of labor" in the sciences, which resulted in a balkanization of knowledge and a splitting off of departments from each other, had created a "pernicious" situation, where a naturalistic culture had made what "everyone knows" is true—that is, materialist Darwinism—something merely accepted on authority, because Darwinists "have so assured the rest of the world." The way to fix this, in Schaefer's view, was for theists to "propose a way out for their secular colleagues."[99] Such theists did not need to be scientists, necessarily. Phillip Johnson, for instance, based his critique of Darwinism on the fact that he was an outsider—he was a lawyer and so was therefore equipped with the tools and training to distinguish rhetoric from evidence. He termed the elites of the scientific establishment the scientific "priesthood," biased in favor of Darwinism because it was the basis of the naturalistic creation story. "The experts," wrote Johnson in the epilogue to *Darwin on Trial*, "therefore have a vested interest in protecting the story, and in imposing rules of reasoning that make it invulnerable."[100] Because of the unified front presented by scientists, "ordinary people thus have no alternative but to accept what the experts tell them about such matters, unless they want to be thought ignorant."[101] Beyond simply worldview preservation, experts were also protective of their own egos. For the astrophysicist Frank Tipler, "the increasing centralization of scientific research has allowed powerful but mediocre scientists to suppress any idea that would diminish their prestige. All great advances in science have by definition the effect of reducing the prestige of the 'experts' in the field in which the advance is made."[102] The very fact that such people were experts—that they had invested years of training and finances into their status—was the very thing that rendered them unreliable. Anti-evolution movements had long made this connection. As the historian Richard Hofstadter (1916–1970) argued, the *Scopes* trial was the first time in the twentieth century that "intellectuals and experts were denounced as enemies by leaders of a large segment of the public." This sort of populist pooh-poohing of expertise—what Hofstadter called the morphing of good, old-fashioned (and often justified) disdain for eggheaded professors into "malign resentment of the intellectual in his capacity as expert"—was central to post-*Dover* ID literature.[103]

But what are experts, and what is their expertise? It would be a mistake, as Hofstadter wrote, to mistake anti-expertise or anti-intellectualism for anti-*intelligence*. The politics of anti-expertise is far more complicated. ID supporters did not think education was worthless (indeed, virtually all of them had advanced degrees). Their hostility to expertise is bound up with their political qualms with liberalism, technocracy, and bureaucracy—especially as manifest in the support for funding and research in the sciences. Such a skeptical stance is not limited to conservative politics either, for the left has its own tradition of anti-expertise, usually manifesting itself in socialist political movements that are hostile to managerial capitalism and the field of economics as a true "dismal science" that props up a rapacious financial system.[104] The enemy in both cases is the bureaucracy, especially those elites who decide the nation's education policy. The Discovery Institute cofounder and former conservative politician Bruce Chapman, in his book *Politicians*, cleanly separated two classes of rulers: politicians (whom he defended against charges of immorality and corruption) and experts (who are the real source of society's problems). Experts—the bureaucrat, technocrats, social engineers—were "glamorized" in the years after Woodrow Wilson and the scientifically managed state of the New Deal era, but they should not be so positively understood.[105] The bureaucrats are parasitic middlemen, and they are often benighted by their own learning, having acquired a kind of ideological blindness and "educated incapacity" (in Herman Kahn's words).[106] The only way to defeat them is the old libertarian chestnut: shrink the government.[107]

"The problem is not expertise," wrote Douglas Axe, William Briggs, and Jay Richards. "We all benefit from experts. The problem is the *tyranny* of experts—when their narrow, professionally biased thinking dictates policy for everyone."[108] This "we know best" approach to politics by unelected but powerful figures is met with particular disdain and can even transform into conspiracy. Axe, Briggs, and Richards continued, many experts are "power grabbing elites" who want control.[109] They cannot be trusted because they cannot admit when they are wrong.[110] Experts "work in silos," which has created an ultracrepidarian habit—they are "cocksure even when they've wandered outside their domain of knowledge." That does not mean that expertise is completely useless, but knowledge is not the same thing as wisdom.[111] Steve Fuller wrote that today's experts are "secular clergy."[112] This does not necessarily reflect a rejection of all science (though it can easily become that) as much as it reflects a deeper suspicion for the edifice of liberal modernity and the way liberalism, science, and technology fused together in the twentieth century through Taylorism and the New Deal state, leading to a huge edifice often called "the establishment," a political-bureaucratic machine that is so powerful that few can resist it and so omnipresent that few even notice it anymore (again, this is not far afield of a left-leaning critique of liberalism, though socialists would never go along with ID's libertarian and pro-capitalist bent). So, what did ID proponents think should be done about this? Expertise, as Axe, Briggs, and Richards wrote, needed to be balanced with "common sense."[113]

The faith in human intuition, the self-evident obviousness of design, and the zealotry of those who resist this conclusion formed the basis of a populist critique of scientific expertise. As Stephen Meyer asked in *Darwin's Doubt*, if ID was so obvious—as it seemed to be, from his perspective—why did experts continually miss the evidence? Meyer's answer was that their metaphysical predisposition, their commitment to materialist science, blinded them.[114] What was unclear to the experts was clear to anyone unfettered by such self-destructive training. For Meyer, it was the everyday experience of design—our witnessing of complex information's only known source: intelligent causation—that formed the basis for its truth. It was "curious," he wrote in *Signature in the Cell*, that we denied "our ordinary intuition about living things."[115] Douglas Axe made the "universal design intuition" central to his argument in 2018's *Undeniable*. It was the "common human faculty by which we intuit design," an intuition that was "reliable when properly used" and that provided "a solid refutation of Darwin's explanation for life."[116] Furthermore, anyone could do this. "If you think these heroes need to have Ph.D.s, I hope to convince you otherwise," wrote Axe; "when it comes to defending the big question of our origin, everyone is scientifically qualified."[117] Grudgingly, Axe admitted that experts do have a role ("even poor arguments might seem to benefit from the status of the people making them"), but he argued that nonexperts were unfortunately mystified by the technical mesmerism fobbed off on them by Darwinists.[118] In response, Axe turned to what he called "common science," which was something like common sense. "All humans are scientists," he stated in the opening section on common science; "basic science is an integral part of how we live. We are all careful observers of our world. We all make mental notes of what we observe. We all use those notes to build conceptual models as needed. And we all continually refine these models as needed. Without doubt, this is science." Unfortunately, "our otherwise trustworthy design intuition must be overruled for the sake of Darwin's theory. But our intuitions aren't easily overruled." In fact, he continued, common science "dispels the elitist myth I [formerly] accepted," and "because everyone practices common science, public reception of scientific claims is arguably the most signified form of peer review."[119] Everyone was an expert when it came to design. "Common experience often counts," wrote Meyer;

"the case for the causal adequacy of intelligent design should be obvious from our ordinary experience."[120]

One can find a grassroots example of this call for common sense in the 2009 self-help book *Game Plan for Life*, by the famed NFL coach Joe Gibbs (and cowritten with the evangelical author Jerry B. Jenkins). Throughout the book, which featured chapters by Gibbs on various life questions that were supplemented with evangelical superstar authors giving additional insight, Gibbs took care to say that he is no one special. Taking delight in the plainness of his first name, he called himself "Average Joe." Turning to God, he wrote, "I think there's more evidence for creation than for evolution." He admitted that he was not an expert but compensated by noting, "I'm just a commonsense kind of guy." The design standbys made sense to him. "I look at the world and nature and things like the human body, particularly the eye, and I wonder how anybody thinks that came about as the result of accident or chance." He also compared life to a wristwatch and the human body to a well-designed machine.[121] Most importantly, Gibbs stressed that his love for his grandchildren could not be simply a biological impulse or mistake, writing, "To say that the ability to love each other and experience the emotion I felt is just an accident of nature makes no sense to me."[122] (Gibbs did not flatly reject experts, however, as he had the Oxford University mathematician John Lennox write the book's remaining section on creation—and Lennox attempted to define creation broadly enough to include YEC and ID supporters).[123] As John Evans argued in *Morals Not Knowledge*, perspectives like Gibbs's are a throwback to the importance of Common Sense Realism to the early fundamentalists. It fit, as George Marsden contended, with American values. One could avoid philosophical experts and confusing terminology by just starting with the dictionary definition of "science." Ordinary people could understand the world if they tried; nature and the Bible are self-evident—he who has ears, let him hear. The danger, their fear, was that common sense would be replaced entirely by arcane and misguided expertise.[124] ID, in taking up this same theme, thereby came as close to the early years of fundamentalist anti-evolution as it ever had.

Years earlier, when David Berlinski wrote his essay "The Deniable Darwin," he took issue with Dawkins's rejection of the design intuition in nature, writing, "It is true that intuition is often wrong—quantum theory is intuition's graveyard. But quantum theory is remote from experience; our intuitions in biology lie closer to the bone."[125] Yet suspicion of the trustworthiness of science would not remain restricted to biology, for many ID advocates doubted or even rejected other scientific concepts. If scientists, the experts, were involved in an ideologically driven campaign of suppression regarding the truth of design and falsity of Darwinism, then it was not implausible to suspect that similarly mischievous cabals were involved in the realm of medicine or physics. Suspicion is a universal acid.

"I have a healthy respect for scientific methodology in its proper sphere," wrote Phillip Johnson in 1996, in what appeared to be familiar drawing of boundaries around science. However, his target was not Darwinism; it was AIDS. "If I were persuaded that the scientific method had been properly employed to determine what AIDS is, how it is caused, and how many people are at risk for AIDS," he wrote, "I would happily accept the judgment of the scientific profession on such matters." This was not the case for Johnson, who doubted that human immunodeficiency virus (HIV) caused the disease known as AIDS (acquired immune deficiency syndrome). "Instead of real science," wrote Johnson, "we have had only HIV-science."[126] Elsewhere, Johnson praised President Thabo Mbeki of South Africa because Mbeki "read the scientific literature, including articles by scientists who dispute the nature of the health crisis that threatens Africa, and he has become skeptical, as most people do when they have an opportunity to study the facts that the official sources do not report."[127]

Nor was Johnson alone in doubting the science consensus outside evolution. William Dembski, who has a son with autism, suggested that vaccines might have caused it (though he admitted he is not certain and that "if vaccines played a causal role, . . . there must have been other predisposing factors"). Dembski was bothered that the film *Vaxxed*, promoted by Robert De Niro, had been pulled from the Tribeca Film Festival in 2016. "'Vaccines are safe and effective.' That's the mantra we continuously hear from Big Pharma," wrote Dembski on his personal website, "as well as the same skeptics who over the years have been implacable foes of ID." He saw some congruence between both causes, continuing, "the pro-vaccine forces see it as a moral mission to suppress discussion of this topic. Andrew Wakefield, the director of *Vaxxed* [and author of the disgraced paper that spawned the antivaccination movement], we are repeatedly told has been thoroughly discredited (discredited by interests incentivized to see him discredited)."[128] Beyond medicine, the ID supporter and journalist Tom Bethell (1936–2021), fellow at the conservative Hoover Institute at Stanford University, authored *The Politically Incorrect Guide to Science* in 2005, a broadside against nearly every mainstream scientific theory. Bethell advocated for intelligent design but also challenged mainstream scientific views on climate change, stem cells, the danger of DDT, and the "political epidemic" of AIDS. Perhaps a bridge too far for publishers of *The Politically Incorrect Guide to Science*, Bethell also challenged the theory of general relativity in a separate work.[129] The Discovery Institute published his rejection of evolution, *Darwin's House of Cards*, in 2016.

The COVID-19 pandemic that swept the globe in early 2020 was a further opportunity for ID proponents not only to double down on the anti-expert rhetoric but also to align themselves with a populist upswing of resentment for the elites who had seemingly been caught flat-footed and then bungled humanity's response. During the pandemic, Douglas Axe, William Briggs, and Jay Richards coauthored a book called *The Price of Panic*, with the clear subtitle *How the Tyranny of Experts Turned the Pandemic into a Catastrophe*. In their reading of events, experts at the Centers for Disease Control and Prevention exaggerated the threat of COVID-19 and pushed the nation into "economic harikari."[130] Anthony Fauci, the former chief medical adviser to the president, was particularly targeted as a "single-minded technocrat." As with Chapman, they contended that the problem was not government per se but was government by expert—especially ones like Fauci.[131] "Globalist technocrats," they contended, "are licking their chops." What was the goal they had in mind? "One-world semi-socialism" shouldn't be laughed off as mere conspiracy.[132] This rhetoric dovetails with the social outlook of the resurgent populism on the US political right, though it differs in still maintaining a broadly libertarian perspective. While there is nothing about intelligent design in such an attack on expertise, it does nevertheless reflect a broad sense of cultural aggrievement at liberal technocracy and the reign of experts. In a nutshell, the question might be, If they're so smart, why is everything so bad?

In 2017, the international affairs specialist Tom Nichols released *The Death of Expertise*, a chronicle of the rapid increase in public animosity toward "experts" in the twenty-first-century US. The dirge of a book covered the way the internet and the accessibility of all information has wrecked the distinction between an expert and amateur and has given the illusion of expertise to many people without the training.[133] He covered the confusion wrought by the modern world and the plaintive cry of the everyday American for something they could understand using only "common sense."[134] Nichols lamented the consequences that followed when experts made public mistakes, which were then exploited by those who sought to discredit the entire enterprise of scientific research—for example, Linus Pauling's quixotic campaign for vitamin C as a cure-all or the 1970s fear that eggs were unhealthy because they contained a lot of cholesterol. Experts had urged Americans to cut eggs from their diet, only to discover later that they were not really a threat and certainly not as unhealthy as the things Americans substituted for them.[135] Nichols feared that these public failures would sow seeds of distrust that could be used against experts in other fields. Jonathan Wells did just that; *Zombie Science* opened with the eggs kerfuffle. "I like eggs for breakfast. . . . I knew I was not supposed to, because some scientists and the U.S. government said they were bad for me," opened Wells with devil-may-care defiance, "but I liked them, and my heart was fine, so I ate them anyway." He noted with satisfaction that in 2015 the government "called off its decades-long War On Eggs," remarking with faux surprise, "'but science said. . . .' Yes, and now 'science says' something else." Be more skeptical, urged Wells to his readers: "You will be told one thing by our enormously powerful and wealthy scientific and educational institutions, as well as by the mainstream news media that serve as their mouthpiece. But you may learn something else if you look at the actual evidence."[136]

The campaign against the establishment, and against scientific expertise, was not confined to the upper crust of design's academic elite, either. Such skepticism found fertile ground in an American public ready to doubt the findings of eggheads everywhere. In 2006, nearly a year after *Dover*, the conservative attorney Andrew Schlafly, son of the conservative activist and critic of feminism Phyllis Schlafly (1924–2016), founded an alternative online encyclopedia titled Conservapedia. Schlafly keyed onto the idea of a conservative encyclopedia when the edits he made to the Wikipedia page on the 2005 Kansas evolution hearing (in which ID played a major role) were frequently deleted by Wikipedia editors.[137] Conservapedia grew to be a large (though not nearly as large as Wikipedia) alternative site for ideologically tinged information on science, complete with pages disputing the dominant views on evolution, climate change, and the theory of relativity. In an interview with Schlafly on National Public Radio, after host Robert Siegel asked him about the Conservapedia page on the kangaroo that explicitly endorsed creationism, he took a postmodern stance on the reliability of knowledge and argued, "It's impossible for an encyclopedia to be neutral. I mean, let's take a point of view, let's disclose that point of view to the reader."[138] The hermeneutic of suspicion has traveled a long way from the radically revolutionary stances of a Freud, a Marx, or a Nietzsche. For Schlafly, an encyclopedia should be tendentious, and any source of knowledge that feigned neutrality was engaging in deception of the self and others.

Intelligent design's campaign against the scientific establishment may have been wedded to a more cerebral style, but its arguments found welcome listeners in the populist world of online political activism. Whereas Wikipedia's entry for intelligent design begins, "Intelligent design (ID) is a pseudoscientific argument for the existence of God, presented by its proponents as 'an evidence-based scientific theory about life's origins,'" the Conservapedia entry (which says "not to be confused with creationism") reads, "Intelligent design (ID) is the empirically testable theory that the natural world shows signs of having been designed by a purposeful, intelligent cause."[139] Conservapedia's articles on evolution link Darwin to both liberalism and atheism, and one notes that "the Americans most likely to believe only in the theory of evolution are liberals."[140] The site's "climate agenda" page (redirected from "climate change") begins with editorial flair: "'Climate change' or climate agenda became the new, ineffective name used by liberals for their global warming hoax, which they coined as it became obvious that there is no crisis in global warming." The article compares the name shift to "what liberals did in redefining 'evolution' to be 'change over time,' which of course is a meaningless expression just as 'climate change' is."[141] Some of ID's leading lights embraced this same kind of suspicion of Wikipedia, too. As Axe, Briggs, and Richards commented in *The Price of Panic*, the real modus operandi of experts (even Wikipedia editors) was the right kind of "virtue signaling." "It doesn't matter whether you think of journalists or tech leaders or Wikipedia editors as intellectuals," they wrote. "What matters is that they think of themselves that way. That's why they align themselves so slavishly with the consensus view of the intellectual establishment."[142]

There is a moral argument here too. As Evans highlights, many Americans do not trust the moral values of scientists (this is possibly a bigger reason for resistance to action on climate change, Evans notes, than mere rejection of scientific evidence or religious opposition).[143] In general, scientists often believe that it is their role in US society, as John Besley and Matthew Nisbet document, "to educate the public so that non-experts will make policy choices in line with the preferences of scientists."[144] Conservatives tend not to share these same values—whether populists or intellectuals. Stephen Meyer wrote that his chief concern with Richard Dawkins was that he was creating a kind of nihilism in the youth.[145] The molecular endocrinologist Christopher Shaw wrote that "the church of scientism has its anti-theist prophets and priesthood who pervade almost all aspects of modern life." These high priests of scientism erode belief in God and open the way for moral collapse because they are no longer bound by the "often inconvenient and lifestyle-intruding rules of a superior being." Instead, "Individuals are free to do as they please, as there is no rational scientific basis for moral concepts such as good and bad."[146] The philosopher J. P. Moreland was concerned that embrace of evolution (even theistic evolution) would undermine the Bible and its teachings that "hell is real and some people will go there" and "that homosexual practice is deeply immoral."[147]

Lurking behind all these beliefs is a far deeper concern than simply the practice of science and the biological mechanism that makes evolution work. It is not, necessarily, even related to religion. As Evans notes, Republican Party affiliation (or at least a conservative moral outlook) is a much stronger predictor for suspicion of scientists than are religious beliefs.[148] As such, politicized as science has become, it is rare that such suspicion of scientists, experts, and even political liberals can be confined to one topic (as we have seen). Writes Evans, "People who do not trust one set of elites usually do not trust any elites."[149]

Toward the end of his testimony at *Dover*, Michael Behe testified that one's preconceptions and metaphysical stance may be the primary driver in one's scientific conclusions (and in this case, he was referring to his own reading of design onto nature). This had been a common refrain from ID advocates since as far back as Phillip Johnson: one's metaphysics determines one's physics. In the case of the populist suspicion of scientific experts and the nature of expertise, along with knowledge itself, one's preconceptions—whether they were political, metaphysical, or religious—were determinative.

There is one area, however, where future overlap and even agreement between ID, theistic evolutionists, and even atheist scientists is possible.

Artificial Intelligence

The Walter Bradley Center for Natural and Artificial Intelligence was officially launched on July 11, 2018. Housed within the Discovery Institute, the center both harks back to the beginning of ID (Walter Bradley being one of the authors of *The Mystery of Life's Origin*) and also its future (tackling a contentious, hot-button topic in artificial intelligence). William Dembski was slated to give the opening remarks at the launch, but because of the declining health of Dembski's mother, Robert J. Marks read them in his stead. The task for which the center set itself, as Dembski outlined, would be to combat the increasingly prevalent idea that machines will attain human intelligence—what is often called Artificial General Intelligence (AGI) or Strong AI. This is more than just an ideological or intellectual exercise. We are at a crucial moment in culture, for those who are suspicious of AI, and the reduction of humanity to machines, their replacement by machines, or their possible subservience to machines must be resisted. "[The center] will not merely demonstrate a qualitative difference between human and machine intelligence," contended Dembski; "it will chart how humans can thrive in a world of increasing automation." Such work should meet a warm reception, but "in an age of computational reductionism inspired by scientific materialism, where so much of the mainstream academy views our humanity as unexceptional and even obsolete, such a vision is anything but."[150] And beyond AGI, the center would also critique and oppose those who herald the coming of a Superintelligence: including those who welcome it, such as Ray Kurzweil, or those who fear it, such as Nick Bostrom.[151] Importantly, the center would not maintain unflagging hostility to AI; rather, "We want to encourage AI's full development, [but] we also want to encourage humanity's full development. . . . The point is to maintain our full humanity in the face of technological progress."[152] As summarized on the center's online landing page, "People know at a fundamental level that they are not machines. But faulty thinking can cause people to assent to views that in their heart of hearts they know to be untrue." The goal, then, would be to help both individuals and society "to realize that we are not machines while at the same time helping to put machines (especially computers and AI) in proper perspective."[153] Further efforts to promote the center have taken the form of a biography of Walter Bradley (cowritten by Dembski and Marks) and the launching of a website, MindMatters.ai, to lay out the ID perspective on the topic.[154] Robert J. Marks, professor of engineering at Baylor, currently serves as the center's director.

The history of artificial intelligence has been somewhat topsy-turvy. The field emerged in earnest in the post–World War II period, but periodic bursts of excitement have been followed by collapses in both public interest and funding. Two fallow periods often called the "AI Winters" diminished enthusiasm for the field. Things changed, however, in the late aughts and early teens of the twenty-first century and for two reasons: the widespread production of Nvidia's GPUs (graphics processing units, usually for high-end PC gaming) and the explosion of massive data sets on the internet.[155] These developments made possible an old but difficult-to-utilize form of AI called "machine learning" (as compared to the other main branch of "knowledge-based," hand-coded AI, or what is sometimes called GOFAI, for "Good Old Fashioned Artificial Intelligence").[156] In the words of Kai-Fu Lee, the former president of Google China, "In the sixty-five years of AI history, arguably there has been just a single breakthrough: deep learning."[157] This is overstated, and notable figures in the field (especially the cognitive scientist Gary Marcus) would strongly disagree. But that said, deep learning has been critical for the explosive growth of AI. Briefly defined, deep learning is a subset of machine learning (with which it is sometimes incorrectly used as a synonym) and is "a type of AI that uses neural networks, which mimic the human brain" (the "deep" refers to the multiple layers in the deep neural network).[158] Mimicry, importantly, does not mean that it is the same as a human brain. As Melanie Mitchell—professor of complexity at the Santa Fe Institute—writes, AI does not "learn" like we do. Instead, "*learning* in neural networks simply consists in gradually modifying the weights on connections so that each output's error gets as close to 0 as possible on all training examples."[159] Recent media hype about AI has often given the wrong impression. "Contrary to what some media have reported," writes Mitchell, "the learning process of [convolutional neural networks] is not very humanlike." Not only do they not learn "on their own," but there is also a collection of "hyperparameters" that "refers to all the aspects of the network that need to be set up by humans to allow learning to even begin."[160] The generative output of these neural networks, however, has been stunningly successful (and has induced a bit of a panic for humanists everywhere), but it is important to keep Mitchell's point in mind. For all its wild success, AI cannot, as of now, learn by abstraction, and it has no "common sense" (an important point for the ID critique).[161] For this reason, the sci-fi writer Ted Chiang notes that it would be better if the term "artificial intelligence" was not used at all and instead we called it "applied statistics," which would be a more accurate term but one that is unfortunately "not as sexy."[162]

These criticisms from people inside and outside the field notwithstanding, the enormous impact and potential of AI in the business world is undeniable. The years after 2011 proved a watershed for AI, with deep learning proving highly lucrative for oligarchic tech companies. Google, for instance, deployed it for search and targeted advertising, as well as Google Translate; Facebook used it for its news feed; and, of course, it erupted into public consciousness with the arrival of ChatGPT.[163] ChatGPT's upending of education, journalism, and the entertainment industry has caused a public reckoning, but AI has been a part of everyone's daily life for years. In focusing on AI before its 2023 irruption into the public scene, the Walter Bradley Center was ahead of the curve. In fact, some of the critical writing it produced on AI would find far-greater institutional agreement than intelligent design ever did (and, in fact, it mirrors many of the criticisms referenced earlier).

Robert J. Marks II, the center's director, is a computer scientist, electrical engineer, and Baylor professor, and he has focused much of his work on AI. Along with George Gilder (one of the Discovery Institute's founders and also a longtime "technology prophet"), Marks wrote that "those who believe AI will exceed the intelligence of humans suffer from a 'materialist superstition' that assumes the mind is a meat computer and can therefore be replicated by AI."[164] William Dembski likewise faulted a priori materialism (and computational theories of mind) as the reason so many people, even experts, get hoodwinked into thinking that AGI is on the horizon (or even possible). Regarding Kurzweil's hope in technological immortality, Dembski wrote, "It is a measure of how deeply entrenched materialism is in Western culture that such possibilities are taken seriously. As someone who has followed the unfulfilled promises of artificial intelligence for more than three decades, I see Kurzweil's transhumanism as closer to a Grimm's fairy tale than to sober science."[165] The equation of humanity to machines, and the mind to computers, has long been part of the religious critique of materialism, but it is interesting to note here that it was ID, after all, that was accused of being mechanistic by its theistic evolutionary foes.

Beyond materialist presuppositions, however, a credulous media is reason for inflated hopes about AI. Identifying media hype as a culprit, Marks argued that GPT-3, for instance, is more style than substance. It is a next-word predictor, is dependent on human-prompt engineering to be original, and is "coherent except when it's not. . . . This may sound spooky, but the closer you look at GPT-3, the less scary it becomes. Calmer, more considered analysis has exposed some weaknesses."[166] Some of Marks's critiques of AI, moreover, deploy similar language to the ID critique of evolution, namely, that a key component of human thought is common sense and the capacity for abductive reason (the "inference to the best explanation" that is central to Stephen Meyer's argument against Darwinism). While he granted that AI could possibly simulate such thinking, it will understand neither "the underlying ambiguity nor the reasons for its resolution."[167] Marks's book *Non-Computable You* was published by the Discovery Institute, but because his line of argument is broadly similar to what non-ID critics have written about AI, it is perhaps not surprising that some ID-affiliated thinkers have published books on AI for major university presses. It is here, unlike most of ID's arguments, that they have found reception outside their usual environment.

In *The Myth of Artificial Intelligence*, the Discovery Institute fellow Erik J. Larson argued that overblown hype about AI actually compromised the ability of the field to make progress. The myth, he wrote, is not that general AI is impossible but that we are on the path toward it—we are manifestly not, he argued. Furthermore, our mistaken belief that we are on the verge of AGI has the side effect of denigrating human intelligence and ensuring "the creep of a machine-land, where genuine invention is sidelined in favor of futuristic talk advocating current approaches, often from entrenched interests."[168] Larson's book was published in 2021 by Belknap, an imprint of Harvard University Press. Though he is affiliated with ID through the Discovery Institute, Larson's publication through a major university press indicates common purchase with some of the ID arguments and mainstream criticism of AI. Larson, like Marks, focused on AI's incapacity for common sense and abductive reasoning. He took up Charles Sanders Peirce's belief that abductive reason is behind most human thinking (recall Meyer's use of Peirce for ID). Machine learning, on the other hand, is basically inductive, and deep neural networks—quoting Yoshua Bengio—"tend to learn statistical regularities in the dataset rather than higher-level abstract concepts."[169] Abduction itself is actually a "weak form of inference" and is often wrong—being, after all, an inference from a single fact to a quick hypothesis (such as why the driveway outside might be wet: because it rained). For Peirce, it could be symbolized as "broken deduction" and even a kind of fallacy, like affirming the consequent.[170] It is, however, critical to human thinking. Peirce wrote, "Deduction proves that something must be; Induction shows that something actually is operative; Abduction merely suggests that something may be." Larson commented, however, that "it's the *may* be—the abduction—that sparks thinking in real-world environments."[171] This language is similar to the kind deployed by Meyer and Douglas Axe regarding evolution. It is our common sense, our abductive capacity—what makes us uniquely human—that shows evolution's problems. Take that away and you get materialism, which might—in the end—dupe us with the false dawn of AGI research. To be clear, Larson's argument does not have anything to do with intelligent design. He did not base his critique of AI on it. But what this shows is that, when intelligent design is removed from the conversation, those who are affiliated with it can find reception for their ideas, even ones that are critical of materialism.

The situation is similar for Gary Smith, an economist at Pomona College and senior fellow at the Walter Bradley Center. Smith wrote two books critical of AI, *The AI Delusion* (2018) and *Distrust* (2023), both for Oxford University Press. Interestingly, Smith tackled questions of scientific expertise but did so in a way that many scientists would probably agree with. Focusing not on the supposed exclusion of design, Smith instead targeted problems in the way science is done—failures like data torturing, p-hacking, and the frequency with which scientific papers are being retracted. He wrote in *The AI Delusion*, "The pressure to obtain publishable results by any means possible is why so much published research is tenuous, flimsy, or outright junk."[172] This makes the public justifiably skeptical, which for Smith was a problem. Too many paper retractions erode confidence in science, and it increases the public sense that science is primarily political rather than intellectual. "The idea that scientists are a tool used by the ruling class to control us has real consequences," he later wrote.[173] In his book *Distrust*, Smith lamented that science's achievements (especially its technological achievements, like the internet) have undercut its credibility. "Ironically," the book opens, "science's hard-won reputation is being undermined by tools invented by scientists. Disinformation is spread by the Internet that scientists created. Data torturing is driven by scientists' insistence on empirical evidence. Data mining is fueled by the big data and powerful computers that scientists created."[174] Like Larson's book, there is no intelligent design in this tome (Smith even positively mentioned Darwin in passing).[175]

Smith's bemoaning of the state of scientific expertise is not the same as the ID jeremiads about persecution discussed earlier. His stance is more akin to Tom Nichols's. The same goes for Erik Larson, who was concerned that "the ethos of Big Data AI is now firmly entrenched in science and culture generally." This ethos creates, then, a sense of distrust in humanity, such that even the future is seen as anti-human. As Larson concluded, "The very point of the myth [of AI] is that anti-humanism is the future; it's baked into the march of existing technology."[176] Perhaps surprisingly, explicit ID proponents have defended expertise in this context too. Marks, in particular, criticized celebrity-driven fears about near-term AGI, writing that figures such as Elon Musk, Bill Gates, and Stephen Hawking were not experts in AI (however much experience they have in business, computers, and science), and so their fears about oncoming AI overlords should not be taken at their word. These figures all spoke outside their "silos of expertise." In fact, this is a place where experts are needed more than ever. The dangers of AI, wrote Marks, could be lessened by "domain expertise," meaning that "AI software should be developed by those with experiential knowledge of the problem being solved. Experts will better identify undesired contingencies during development of the AI software."[177] The shoe was now on the other foot regarding expertise. Nonexperts might be needed to unseat Darwin, but experts were needed to combat media hype and bad ideas in AI.

Unlike the wrangling over Darwin, this is an area where the ID affiliated can find agreement with other academics and, in fact, where they might be moving closer to more mainstream positions. For one thing, the concern about maintaining humanity in the face of machines finds august precedent in the work of legendary critics of technology—among them Lewis Mumford and Jacques Ellul—especially comments like Dembski's: "the point is to show society a positive way forward in adapting to machines, putting machines in service of rather than contrary to humanity's higher aspirations."[178] But people with immediate concerns about AI can find common ground too. Even those who are less techno-pessimistic than classic philosophers of technology would be inclined to agree. Gary Marcus, for instance, a persistent gadfly in the AI world, has long argued that the current pursuit of deep learning AI as the be-all and end-all of artificial intelligence is misguided and that "deep learning is hitting a wall."[179] He has instead advocated for a return (in part) to "GOFAI" approaches that could be combined with machine learning. The only way to achieve common sense, general intelligence, and "deep understanding" is to "look . . . for a compromise."[180] This is, in his view, the only way to get AI that we can trust is good for humanity.

The point, in concluding this chapter on artificial intelligence—which seems at times to be removed from intelligent design—is to show that there are future-oriented parts of the ID movement that might find not only agreement but also support from outside the ID world. Design arguments failed to carry the day, and ID's opponents—both atheistic scientists and theistic evolutionists—have remained as steadfast as ever. But some of the language of design—the importance of common sense, abductive reasoning, and the special capacity of humanity—has found resonance outside ID, and common cause could be made here with those very same antagonists. While many critics of the machine analogy in nature, and of computational theories of mind, would probably still hold that ID is as mechanistic as its materialist opponents (Edward Feser, for instance, the Thomist critic of ID, has criticized computationalism at length), it is possible that their agreement on AI could overshadow their disagreement on whether cells are full of literal machinery.[181] One could speculate that in AI criticism, ID might have a more vibrant future with more diverse allies, but it would have to sideline many of its past focuses—including perhaps the machine analogy on which the design argument often depends.

Conclusion

In *Religion vs. Science: What Religious People Really Think*, the sociologists Elaine Howard Ecklund and Christopher Scheitle note that there is no necessary link between different forms of science skepticism; that is, there is nothing inherently present in anti-Darwinism to logically compel a skeptic to also reject, say, the consensus on climate change. The real issue is not the understanding of science but the politicization of the circumstances around climate change, leading certain groups to reject it no matter what.[182] It is hard not to see the same sort of political climate—a hermeneutic of deep suspicion—surrounding intelligent design in its post-*Dover* years. While there have still been efforts at mainstream, or at least mainstream-like, scientific enterprise—in the establishment of journals, the publication of findings, the funding of lab work (as with BioLogic)—there has been a far stronger strand of defiance, anti-establishment rhetoric, and the denunciation of experts qua experts. The doubling down on previous arguments in the face of legal defeats gave the impression that design advocates saw that ID struggled to succeed not because it was wrong but because it was unfairly targeted—the institutions were biased against it, the playing field was not level. The experts were the ones who stacked the deck, and if the establishment was innately and universally hostile to design—such that the scientific world and the media in its thrall would array themselves against it—then why would experts not do the same kind of thing in their conflict with dissenters over vaccines, AIDS, climate change, or even general relativity? The one major outlier here, artificial intelligence, is not enough to counter what appears to be a dominant trend.

The archetypal figure of this, in the story of design, might be William Buckingham, the instigator of the *Dover* trial and intransigent witness on the stand at the end. As Edward Humes notes, Buckingham's performance was a microcosm for larger cultural trends, regarding not only science but expertise as such. Humes's portrait was both troubling and sympathetic—here was a man of strong beliefs, a man who was convinced of the truth and rightness of his actions, a man who could not easily change his mind because "his beliefs are too much a part of who he is, and mere facts do not easily dent them." He was a man whose faith saved him from a brutal workplace accident, from the pain and suffering of a lifelong physical injury and its corresponding painkiller addiction that it wrought on him, a man who came close to suicide in desperation for release from this vale of tears and who only was rescued by the thin life preserver offered him by a neighboring church. He had reasons for doing what he did, and in this lies the moral drama of *Dover*, of intelligent design, and the agonizing slide into universal suspicion that characterizes contemporary political life. Intelligent design may not itself have invented such widespread doubts, focused initially as it was on a narrow scientific question—the efficacy of natural selection acting on random mutation—but it helped sustain and intensified conditions in which Buckingham could thrive. His rejection of evolution "would be no one's business or concern but his own, except that those beliefs found their way into the business of the Dover Area School District. And then they became everyone's business."[183] That business is unresolved—perhaps it always will be—but the story of William Buckingham is the story of the United States; the story of evolution, Darwin, and design; and the story of politics, pandas, and people.

Conclusion

Design and the Future

In 2016, William Dembski abruptly retired from intelligent design. After the release of his 2014 book *Being as Communion*, he surveyed his career and then told the apologist Sean McDowell, "I'm not sure I have a whole lot more to add." His seemed less optimistic about ID's future. "In the long run," he said, "I do see ID as succeeding. But as John Maynard Keynes put it, 'in the long run we're all dead.' I don't have a crystal ball, and I'm not holding my breath that we're going to see ID victorious, as in becoming the dominant paradigm of biological origins, any time soon."[1] ID's supporters were surprised by the admission, but Dembski officially confirmed on his personal website that he was retired. "The camaraderie I once experienced with colleagues and friends in the movement has largely dwindled," he explained. "I'm not talking about any falling out. It's simply that my life and interests have moved on." He did take care to note that this did not mean that he renounced ID or its arguments.[2] Following this departure, Dembski focused on some related matters, including a biography of Walter Bradley, but ventured away from ID in cowriting a book on the legendary baseball player Steve Dalkowski, who threw harder than anyone but could not refrain from issuing walks.[3]

Dembski's departure was not the only blow dealt to ID. In 2019, Phillip Johnson died at his home in Berkeley. In a retrospective, David Klinghoffer wrote that Johnson had "lit the match" of ID and that by shifting the conflict over evolution to naturalism and not the Bible, he had "transformed the entire origins debate."[4] John Mark Reynolds noted mournfully that his mentor's death could be summarized as "Gandalf has gone into the West."[5] By 2020, two of ID's most prominent voices—Dembski and Johnson—were no longer a part of the movement.

In Klinghoffer's memoriam, he argued that Johnson's contribution was framing the origins argument around philosophy and metaphysics.

As we have seen, ID emerged and departed from creationism—partly in response to the 1980s court cases and development within scientific creationism, partly because of new directions in the fields of biochemistry and molecular biology, and partly because of the ideological differences among its advocates—and it jettisoned much of what made young-Earth creationism distinctive: biblical literalism, Noah's Ark, and anti-uniformitarianism. In place of the Bible, the flood, and the young Earth, ID proponents constructed a worldview based on a philosophical postulate of theism and a self-described metaphysical openness to divine causation in nature. They rescued the design argument from its natural theological and Paleyite heyday, and they reframed it not as a theological principle but rather as an abductive inference to the best explanation. In softening or even removing the biblical language about creation, they were able to build a movement that was—from their view—scientifically and not religiously based. No dogmas, catechisms, or divine revelation would influence the design worldview. As this book has shown, such a minimalist theological foundation meant that ID was open to political and social collaboration with anti-evolutionists of all stripes, thus enabling the ID movement to become a "big tent," under whose auspices young-Earth creationists, old-Earth creationists, wayward theistic evolutionists, and agnostic rabble-rousers could find a home. This coalition, unified more by a conservative political ethos and a hostility to Darwin than by a clear theological or religious platform, enabled ID to achieve a rapid degree of success in the public sphere, infiltrating and modifying science curricula around the United States in the late '90s and early 2000s. In Dover, Pennsylvania, however, ID was undone by its own cultural and political successes, and by the populist wave of anti-evolutionary support it tried to tame, as it was pressed into defending itself in a court case that its leaders had not wanted. ID was humiliated in *Dover* by the testimonies of its opponents, the extensive cross-examination of its own witnesses, the mendacious behavior of its creationist supporters, and the uncompromising ruling of Judge Jones. In the years since, ID proponents have doubled and tripled down on the same arguments they made before the *Dover* case: irreducible complexity, icons of evolution, or the (in)ability of Darwinian mechanisms to generate the first cell. Coupled with this strict maintenance of their ID framework was a rising sense of cultural aggrievement and unjust persecution, a siege mentality that had always been present but that roared to the forefront and became the dominant form of ID rhetoric post-*Dover*.

Data from Google searches tell an interesting story about the waning public attention given to ID. Peak searches for intelligent design were in 2005, during and shortly after the *Dover* trial; since then, searches for ID have fallen off dramatically, indicating a severe drop in outsider interest. Some brief spikes coincided in 2008 with the release of *Expelled* (and possibly the 2007 Creation Museum), but searches hovered around 1–3 percent of the 2005 peak from 2015 to 2024. Unlike ID, creationism continued to garner interest past 2005 (though it too appears to be diminishing by the early 2020s). It is important not to draw too general of a conclusion from this—Google searches are only one small metric by which to judge a movement's vitality. The dedication of the people within these movements has not waned, but it does possibly indicate a drop in attention from people outside the subcultures of anti-evolutionism. The most recent spike was a surge of interest in young-Earth creationism probably caused by the highly publicized debate between Ken Ham and the science popularizer Bill Nye, in January 2014. Casey Luskin lamented what he saw as the debate's harmful impact, writing on ID's *Evolution News* that it gave mainstream audiences the impression that anti-evolution necessitated a young-Earth worldview, with insufficient communication that the "mainstream scientific view about the age of the earth is *totally compatible* with an intelligent design view that *totally refutes* Nye's intolerant, materialist beliefs about the history of life."[6] Ham announced that his debate with Nye spurred enough public and financial interest in creationism that his organization, Answers in Genesis (AiG), had the resources in 2016 to launch a new hybrid museum/theme park dedicated to Noah's Ark, called Ark Encounter (a different museum from AiG's Creation Museum, both in Kentucky).[7]

The world of anti-evolution after *Dover* was in some respect a parting of the ways between ID and young-Earth creationism. Though the two styles of anti-evolution were ideologically compatible in a minimalist sense (and, as chronicled earlier, ID's separation from YEC was occasioned by its discarding of most of the latter's distinctive features), the social confluence of creationism and ID that led to grassroots support across the country broke down in the aftermath of the 2005 court failure. As of the early 2020s, there was as much mutual antagonism between

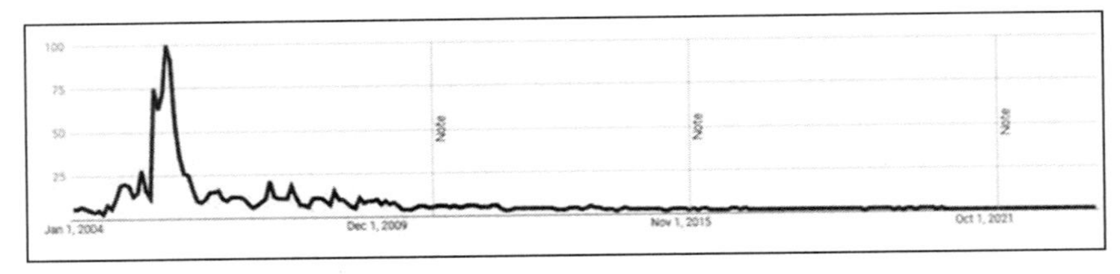

Figure C.1. Google search interest for the phrase "intelligent design." Peak searches were in 2005, with a few short bursts in 2007–2008 and only 1–3 percent interest compared to peak levels since 2014. Since 2021, searches average at 1 percent of peak interest. (Google Trends, December 19, 2023)

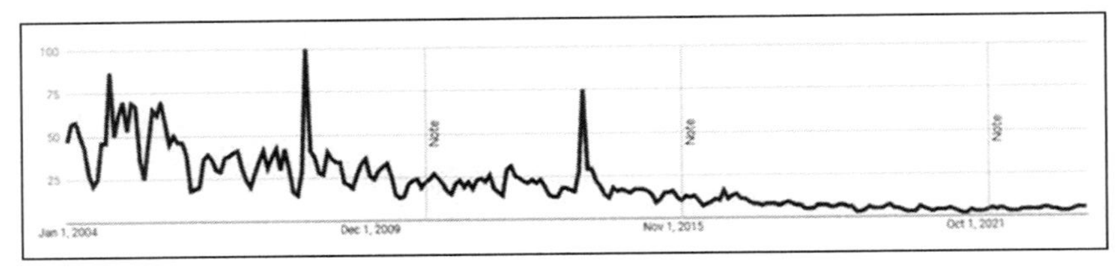

Figure C.2. Google searches for "creationism" have been more frequent than intelligent design and have managed to peak in 2008 and 2014. However, since 2014 and the Ham-Nye debate, search interest has steadily declined. (Google Trends, December 19, 2023)

IDers and young-Earth creationists as ever. Dembski described the frayed relationship in the McDowell interview: "[YECs] were friendly to ID in the early 2000s, until they realized that ID was not going to serve as a stalking horse for their literalistic interpretation of Genesis. After that, the young-earth community largely turned away from ID, if not overtly, then by essentially downplaying ID in favor of anything that supported a young earth." The worst result was the Ark Encounter. "The Noah's Ark theme park in Kentucky is a case in point," he said. "What an embarrassment and waste of money."[8] Nor was this merely a matter of abstract disagreement. Dembski's career was hampered by accusations of heresy when he published *The End of Christianity*, a book that took seriously the problem of evil posed by the existence of death before the Fall of Man. In a 2019 interview with James Barham, Dembski explained, "Ken Ham went ballistic over [the book], going around the country denouncing me as a heretic, and encouraging people to write to my theological employers to see to it that I get fired for the views I take in it." Dembski's friends, he recounted, used to joke that his politics "put [him] to the right of Attila the Hun," but in the twenty-first century, that might still be too liberal. He lamented the fact that, in conservative Christian circles, "one can never be quite conservative enough." He had thought—naively, he admitted—that *The End of Christianity* might move some young-Earth creationists away from flood geology and toward ID. "I couldn't have been more wrong," he concluded.[9] Ken Ham and AiG spurned ID, and the attitudes toward intelligent design at both the Creation Museum and Ark Encounter range from cautiously suspicious to outright hostile. ID-*ish* language is deployed a great deal in the museum, especially in the Wonders of Creation room, which focuses on the beauty, precision, and complexity of nature, but there is very little explicitly related to the intelligent design movement and nothing regarding its primary arguments, such as irreducible complexity. The design language present is vague and imprecise. As Susan and William Trollinger note in their book on the Creation Museum, the museum occasionally features "the language of ID," but in general, "AiG seems to resist the intelligent design movement on the grounds that it 'fails to reference the God of the Bible and the Curse's impact on a once-perfect world.'" To avoid giving the impression that ID and young-Earth creationism are the same, the Trollingers opted to use the phrase "divine designer" to describe the AiG deity, rather than "intelligent designer."[10] Furthermore, James Bielo, in his book on the Ark Encounter, recounted that the team behind it viewed ID supporters as a skeptical target audience to be evangelized, along with evolutionists, secularists, and nonbelievers.[11]

It is difficult to see how ID might be able to compete, in the twenty-first century, with the style, money, and entertainment value of the Creation Museum or Ark Encounter. Of what use is specified complexity against a zip line, petting zoo, food court, and planetarium? Bielo, in *Ark Encounter*, argued that the AiG brand of creationism may be better suited for the social-media-tinged, entertainment-saturated, postmodern world. The arcane nature of ID's empirical arguments against Darwin ultimately do not carry the same pizzazz as sociopolitical campaigns against evolution or liberal immorality. "Religion-science need not persist as our only analytical frame for understanding creationist identity and ambition," wrote Bielo, arguing instead for a "recalibration" that frames "religion-entertainment [as] the center of analytical gravity." The Creation Museum and Ark Encounter are both museums and

Figure c.3. The Wonder of Creation room at the Creation Museum. The room is filled with reverence for divine design but evinces little overt connection to ID concepts. (Photo by author)

Figure C.4. A wrecking ball of "millions of years" destroying the side of a church. The image is meant to symbolize the destructive effect of the old Earth, a view adopted by liberal churches (and by extension IDers) before their cultural demise. (Photo by author)

theme parks. Such places, rather than a think tank like the Discovery Institute, may be the future hub of anti-evolution.[12] While ID shares with young-Earth creationism a presuppositionalist approach to scientific evidence (with one's religious—or irreligious—starting point determining one's view of nature), it may ultimately be too much of an old-school enterprise to compete in the future. Too wedded to a cerebral, academic approach to argument, ID may have committed the only twenty-first-century crime worse than being incorrect: being out of date. Campaigns against established scientific knowledge have not abated, though, and ID continues to draw interest from the professionally suspicious, as Stephen Meyer's 2023 appearance on *The Joe Rogan Experience* podcast indicates. Rogan is not an intelligent design advocate, but the platforming of Meyer is significant in science perception around the world. In one excerpt that Rogan uploaded to YouTube, Meyer focused on defending the Big Bang theory, after which Rogan neatly summarized (borrowing from Terrence McKenna) what could be taken as the core ID position on God and science in general: "Science wants you to believe that it's all about measurement and reason, if you allow them one miracle. That one miracle is the Big Bang. That all things come from the most preposterous idea ever—that everything came from nothing and one big miracle." Meyer replied, "I completely agree. I love that."[13]

ID's lasting contribution thus looks to be not its challenge to Darwin but its post-*Dover* challenge to the reliability of scientific practice and its both tacit and explicit support for a host of suspicious movements that are ideologically, if not practically, unmoored from any religious worldview: vaccine hesitancy, AIDS denialism, and climate change denial.[14] It remains to be seen if the potentially broader compatibility of ID's AI criticism will counter this trend.

An Important Legal Caveat

One crucial legal development must be scrutinized as we conclude. Norman Geisler, at the end of his book *Creation and the Courts*, despaired of the lack of conservative influence on the Supreme Court. To help creationism win, he wrote, people needed to vote for conservative politicians. It would be difficult to penetrate the "virtual liberal monopoly" in the courts—from the district to Supreme Court level—but it "must become a number one priority if we are ever to restore the teaching of creation in our public schools." It looked like that would be very difficult at the time of his writing (2007), because there was a liberal majority and "no immediate prospects for getting another conservative Supreme Court justice."[15] Things have changed quite a bit since then. By the early 2020s, the makeup of the Supreme Court was decidedly conservative, and a 2022 court ruling has raised questions about the future of creationism in schools. The *Kennedy v. Bremington School District* decision revolved around the high school football coach Joseph Kennedy and his on-field prayers after the conclusion of his team's games. The resulting decision in his favor may possibly have enormous effects on future cases. In the dissenting opinion, Justice Sonya Sotomayor (along with Stephen Breyer and Elena Kagan) wrote, "Today's decision goes beyond merely misreading the record. The Court overrules *Lemon v. Kurtzman* . . . and calls into question decades of subsequent precedents that it deems 'offshoot[s]' of that decision."[16] The *Lemon* test, which stemmed from *Lemon v. Kurtzman*, was critical to the defeat of creationism in the '80s and ID at *Dover.* Effectively overturned, it is unclear what this will this mean for a hypothetical future Supreme Court case on the legality of teaching creationism or intelligent design in public schools. Will the legal future of these anti-evolutionary movements look different than their past?

Perhaps; perhaps not. Prognostication is a foolhardy endeavor, especially for the historian. As G. K. Chesterton (1874–1936) observed, one of humanity's greatest pastimes is a game called "Cheat the Prophet," in which "the players listen very carefully and respectfully to all that the clever men have to say about what is to happen in the next generation. The players then wait until all the clever men are dead, and bury them nicely. They then go and do something else."[17] What the future holds for ID and for anti-evolution movements more broadly is something God only knows (in fact, in a perfect example highlighting the difficulty of prediction, Dembski actually semi-unretired from ID in February 2021).[18] That said, one thing that seems certain is that the suspicion of science, the sense of aggrievement, persecution, and intolerance, and the sociopolitical world of resentment and betrayal that ID legitimated with its rhetoric and—ironically—the credentials of its supporters will persist. Particularly during the COVID-19 pandemic, it was difficult not to see in the opposition to federal science officials, popular resistance to mask mandates, the conspiracies regarding 5G networks, and the politicization of nearly everything the same sort of populist, antiestablishment mass movement that enabled ID to thrive so successfully. ID may not have been equipped to survive in such a world—it was too dedicated to twentieth-century conservatism's hierarchical view of political order, too focused on absolute moral law, and too committed to the reliability of scientific observation of mechanical nature—but it helped support the conditions for other movements to thrive. Despite its elitist political sensibility, ID rode a populist wave that finally crashed at *Dover*, leaving its leaders unable to regain their social and political influence in the aftermath.

The Question of God

God's existence, though, was always the focus of ID, and so one must ask, What about the debate over design itself? The chief thrust here was, at bottom, one about meaning and purpose. Stephen Meyer wrote honestly and with great vulnerability about his personal doubts as a teenager. "The problem of human significance began to torment me," he confessed. He tried to think his way through it, tried to focus on something that he really cared about in baseball, but it did not help. What good was reading about Honus Wagner or Tom Seaver? They would all be dead and forgotten one day. The Hall of Fame is no antidote to mortality. Turning to advice and self-help was no way out, either. "I was a mess," he wrote. "It seemed a dark cloud followed me throughout my day. I remember thinking that, at fourteen, my life was over."[19] ID, for Meyer, meant a worldview that gave purpose and hope not only to him but to the rest of humanity, too. "The return of the God hypothesis," he concluded, "also revives a hopeful possibility—that our search for ultimate meaning need not end in vain."[20] There is no doubt about his sincerity here, but one does have to ask the question about what ID would mean for ultimate meaning, even if it were proved beyond possible objection.

A powerful thought experiment in this regard came in the form of a short story by Ted Chiang called "Omphalos." Part of his award-winning collection *Exhalation*, the story takes place in a world in which creationism is true. In this world, the church holds incredible power—

both culturally and scientifically. The protagonist, Dr. Dorothea Morrel, is a creationist scientist writing a prayer, because her faith is being tested. While it might seem obvious, in such a world, that Christianity is unassailable fact (and, indeed, there is scientific evidence of young-Earth creationism in the form of mummies without navels), a scientific discovery threatens to destroy the entire edifice of the faith. An astronomy paper will soon be published: a seemingly geocentric planet has been found elsewhere, a planet locked in the center of its galaxy with its sun and stars orbiting it rather than the other way around. "The significance of the miracle is the real question," muses one character. "What does this marvel tell us about God's design?" The plausible answer is unsettling. Earth, in fact, might not be that important. This geocentric planet, and its possible inhabitants, must be—for the story's characters—"the reason God created the universe." Such a shattering truth upends Dr. Morrel's entire worldview. Like Stephen Meyer, she saw the world change from one that had previously been imbued with meaning to one now devoid of it. Humanity is an afterthought, possibly an accident—even if we were created by God. As one character asks, "What if God had no intentions about us at all?" After all, Dr. Morrel speculates, should not humanity be the center of the universe, if we are the most important beings? She infers to the best explanation: "Our solar system should be the fixed point against which all else is moving; our Sun should be at absolute rest. If the evidence doesn't support that premise, then we must ask where our commitment truly lies." Dr. Morrel obsesses over the suddenly terrifying ordeal creation must have been. She had assumed that "the primordial humans knew from the moment they drew breath: that they were created for a reason." "The possibility that they didn't know this," she reflects, "is something I haven't been able to stop thinking about." Consider how frightening it would be to simply pop into existence with no clue as to why or how you appeared. "I've tried to imagine," she writes, "what it must have been like to awaken fully formed, possessing certain skills but without a past to remember, lost in a world of amnesiacs. It seems terrifying to me, even more terrifying than what I've experienced these last few weeks." But, in the end, she has no choice but to soldier on. "Even if humanity is not the reason for which the universe was made," she concludes, "I still wish to understand the way it operates."[21]

Importantly, though, God's real purposes are never revealed in the story. All of these ideas, even the terrible truth, are at bottom assumptions from the creationist scientists about what the world *should* look like, in their minds, if God had created it. As Chiang writes in his appendix notes, "It seemed to me that that if humanity really were the reason the universe was made, then relativity shouldn't be true; physics should behave differently in different situations, and that should be detectable."[22] The irony, then, is that, for Chiang, the confusing, nonstandard world in his story is what it would look like if God *had* created it for humanity—the author and the characters are at odds. They thought they knew what a world created by God should be like, and when it did not conform to what they expected, their faith unraveled. From a limited data set—the strange movement of a previously unidentified planet—the entire foundation crumbled. But the opposite truth is just as important: even with so much evidence—as obvious as trees without rings, sediment that shows the Earth is young, mummies with no "omphalos" (belly buttons)—this story beautifully illustrates the way certainty eludes all of us. Doubt always finds room to possibly shipwreck faith. Supposing, then, that intelligent design really were proven—that irreducible complexity was undeniable, for example, or that specified complexity showed without a shadow of a doubt that the information in the first cell must have been created by God—would that alleviate the problems that have always bedeviled faith? The problem of evil? The hiddenness of God? Whether or not Christ truly resurrected? The problem of incompatible revelations and other religions? Would not all of us still be in the flux and confusion that afflicted Dr. Dorothea Morrel?

This is not an easy thing to answer. In the end, though, prediction is a fool's errand, so it may be best to turn to William James (1842–1910) and give him an airing, since he asked a similar question at the end of the nineteenth century and his prediction has been borne out so far. The real meaning of the ageless debate between materialism and theism, he argued, was not about "hair-splitting abstractions about matter's inner essence, or about the metaphysical attributes of God. Materialism means simply the denial that the moral order is eternal, and the cutting off of ultimate hopes; theism means the affirmation of an eternal moral order and the letting loose of hope. Surely here is an issue genuine enough, for anyone who feels it; and, as long as men are men, it will yield matter for serious philosophic debate."[23] Over a century has sustained the accuracy of this prediction, and it seems to me a safe bet that another century will not see any real change. Whether it takes the form of ID, whether creationism continues to be a vibrant rural force, or whether theistic evolution can overtake its competitors and prove to be the fittest theological survivor remains to be seen. But the debate over God and science, religion and atheism, hope and meaning will almost certainly never cease.

ACKNOWLEDGMENTS

No book is an island, and this one wouldn't exist without the contributions and support of many people. In its first incarnation, crucial first steps were encouraged by Grant Wacker, who years later also provided generous feedback on the final product. Duke professors Kate Bowler, Joseph Winters, and Mohsen Kadivar all graciously read and reviewed it. But the most help came from my *doktorvater* David Morgan, who read, marked, and commented on every draft of every chapter with blinding speed, reliability, and insight.

I wish I could thank the late Ron Numbers again for his encouragement and guidance throughout this book's life. There is no way it would have either been started or finished without his mentorship.

I would like to thank the reviewers who contributed their insight and suggestions as the book evolved—both those anonymous and those named. Especially helpful was Francis Beckwith, who not only read the project twice but provided two dozen pages of notes and recommendations.

As the book took final form, I am grateful to Jennifer Hammer for her editorial guidance, patience, and ability to shepherd me to the finish line.

I owe a great debt to my friend Zachary Zschaechner for being the first one to suggest I abandon an early project and focus instead on intelligent design (something of a full-circle experience, since we became friends fifteen years ago at UC Riverside during a class conversation about Christianity, science, and ID). I also received great feedback from my writing group compatriots Steven and Becky Elmore, Brittney Stoneburg, Michael Sutherlin, and Manuel Montaño, who read chapter drafts at various stages.

Lastly, I would not have gotten anywhere without family. My parents, Scott and Karen Howell, to whom this book is dedicated, are the ones who taught me to love learning, reading, and writing. And my wife, Mary Carol, makes everything possible with her unflagging love and support.

NOTES

INTRODUCTION

1 Schaefer, foreword to *Mere Creation*, 9.
2 Denton would outline his more directed model of evolution in his later book *Nature's Destiny*.
3 Campbell, "Report on the Mere Christian Conference."
4 Woodward, *Doubts About Darwin*, 171–172.
5 Klinghoffer, "Mere Manipulation."
6 Cootsona, *Mere Science and Christian Faith*.
7 Dembski, introduction to *Mere Creation*, 29.
8 Johnson felt that the 1960 film *Inherit the Wind* created a retroactive false memory of the *Scopes* trial in Tennessee. Historians generally agree that it is not a good source of information about the 1925 trial—in which the Tennessee schoolteacher John T. Scopes was charged with illegally teaching evolution. The stereotype that annoyed Johnson was that that all those who objected to evolution were simply backward and ignorant Bible-thumpers, easy fodder for cruel journalists and the national media.
9 Johnson, "How to Sink a Battleship," 453.
10 Ury, "Reports: Mere Creation Conference," 25, 30.
11 "School District Challenges Darwin's Theory"; Zimmerman, "Creationism in New Hampshire."
12 *Kitzmiller v. Dover*, 440 F. Supp. 2d 707 (2005).
13 Rosenhouse, "Twenty Years After Darwin on Trial, ID Is Dead."
14 Klinghoffer, "Is the Market for Articles That Ask 'Is Intelligent Design Dead?' Dead?"
15 Menuge, "Who's Afraid of ID?"
16 Woodward, *Doubts About Darwin*.
17 Woodward, *Darwin Strikes Back*, 10.
18 Caudill, *Intelligently Designed*.
19 Numbers, *Creationists*, 373–398.
20 G. Evans, "Reason or Faith?"
21 Numbers, *Creationists*, 14.
22 Quoted in Keeley, "Art of Poetry XIII: George Seferis," 83.
23 See Bulgakov, *Bride of the Lamb*.

CHAPTER 1. CREATIONISM

1 Webb, "But Will He Check That Recycle Bin Again?"
2 Numbers, *Creationists*, 398.
3 Forrest and Gross, 7, 16.
4 Stenger, "Intelligent Design," 21.
5 Pennock, *Tower of Babel*, 6, 28.
6 Forrest and Gross, *Creationism's Trojan Horse*, 273, 275.
7 Pennock, *Tower of Babel*, 28–31.
8 Roberts, "Sedimentary Muddle."
9 Behe, *Darwin's Black Box*, 5. This admission led Kenneth Miller, a Catholic, pro-evolutionary biologist, to opine in a review, "Creationists who believe that Behe is on their side should proceed with caution. . . . The careful reader will recognize this book for what it truly is—an argument against evolution that concedes nearly all the contested ground to Darwin's edifice." See K. Miller, "*Darwin's Black Box* Reviewed," 40.
10 Dembski, *Intelligent Design*, 247. Dembski discussed scientific creationism briefly in this section, noting that it was characterized primarily by "prior religious commitment." In an interesting mirroring of his own predicament, this was something that scientific creationists often denied as well.
11 Dembski, *End of Christianity*. See also Nettles, "Book Review."
12 Meyer, "Letter from the Ad Hoc Origins Committee," 226.
13 Menuge, "Who's Afraid of ID?," 35.
14 Scott, "Darwin Prosecuted," 36.
15 Forrest, "Still Creationism After All These Years," 190–191.
16 Ham, "Response from Young Earth Creationism," 209.
17 Ham, 210.
18 Morris, "Neocreationism."
19 Geisler, "Review of Michael Behe's *The Edge of Evolution*."
20 Nelson and Reynolds, "Young Earth Creationism." See also Nelson, "Life in the Big Tent."
21 Rios, *After the Monkey Trial*, 9.
22 Numbers, *Darwinism Comes to America*, 21, 49.
23 Darwin, *Origin of Species*, 139; Numbers, *Darwinism Comes to America*, 50.
24 Plantinga, *Warranted Christian Belief*, 245.
25 Marsden, *Fundamentalism and American Culture*, 214–215.
26 Marsden, 57.
27 Marsden, 110–111.
28 Reid, *Lectures in Natural Theology*. This edition of Reid's *Lectures on Natural Theology* was reprinted under the editorial direction of the ID supporters James A. Barham and Jake Akins. See Dembski, "Thomas Reid Lectures on Natural Theology."
29 Marsden, *Fundamentalism and American Culture*, 169.

30 Marsden, 174.
31 Marsden, 213.
32 Quoted in Davis, "Fundamentalism and Folk Science," 221.
33 Davis, 237.
34 Davis, 218.
35 Davis, 238.
36 Geisler, *Creation and the Courts*, 74. For more on Warfield as an evolutionist, see Livingstone and Noll, "B. B. Warfield." Some ID advocates would later dispute this interpretation of Warfield.
37 Edward Larson, *Summer for the Gods*, 225–246.
38 While Sputnik is a memorable symbol of this shift, some historians have argued that this is too simplistic and that there were trends toward reintroducing Darwin to high schools before the satellite's launch. For an example, see John L. Randolph, "Myth 23."
39 Muller, "One Hundred Years Without Darwin Are Enough."
40 The classic overview of the modern synthesis is the aptly titled *Evolution: The Modern Synthesis* by Julian Huxley. For a historical assessment of evolution throughout its history, including the modern synthesis, see Bowler, *Evolution*. The modern synthesis is the dominant view, but it is not the only one, as there are alternative takes on evolution. See Rupke, "Myth 13"; and Pigliucci and Muller, *Evolution*.
41 Bryan, "God and Evolution."
42 A. Wallace, *Darwinism*.
43 "Days Six & Seven," 299, 302.
44 Numbers, *Darwinism Comes to America*, 52.
45 Kazin, *Godly Hero*, 273–274; Numbers, *Creationists*, 56.
46 T. Huxley, "Origin of Species."
47 See Bowler, *Non-Darwinian Revolution*; and Bowler, *Eclipse of Darwinism*.
48 Bateson, *Materials for the Study of Variation*. For a short overview of saltationism and the controversy between Bateson and Darwin's cousin Francis Galton, see Gillem, "Evolution by Jumps."
49 Bryan, *In His Image*.
50 J. Huxley, *Evolution*, 22.
51 For a detailed discussion of the American Scientific Affiliation, see Rios, *After the Monkey Trial*. See also Kalthoff, *Creation and Evolution*.
52 Morris, *History of Modern Creationism*, 74–5.
53 Morris frequently called young-Earth creationism "sound, Biblical" creationism, as opposed to unsound and unbiblical ideas like the local flood, the ancient Earth, and the existence of death and predation before the Fall.
54 Numbers, *Creationists*, 217.
55 Whitcomb and Morris, *Genesis Flood*, 35.
56 Whitcomb and Morris, 131.
57 Dobzhansky, "Nothing in Biology Makes Sense," 127.

58 For a discussion of the legal ramifications and goals of scientific creationism, see Edward Larson, *Trial and Error*, 125–155.
59 Morris, *Scientific Creationism*, iv.
60 For a deeper look at the transition of biblical creationism into scientific creationism, see Numbers, *Creationists*, 268–285.
61 Eldredge, *Triumph of Evolution*, 12.
62 Kitcher, *Abusing Science*, 180.
63 Coyne, "God in the Details," 227.
64 Whitcomb, "Review Article," 294–295.
65 Watson, *Double Helix*, 197.
66 S. Miller, "Production of Amino Acids."
67 Darwin, letter to J. D. Hooker.
68 Suarez-Diaz, "Molecular Evolution in Historical Perspective," 206.
69 Lazcano, "Historical Developments of Origins Research," 13–14.
70 Kenyon and Steinman, *Biochemical Predestination*, ix.
71 Calvin, foreword to *Biochemical Predestination*, vii.
72 Kenyon and Steinman, *Biochemical Predestination*, 30.
73 Kenyon and Steinman, 265.
74 Meyer, *Signature in the Cell*, 229–230.
75 B. Wilder-Smith, preface to *Let Us Reason*.
76 A. Wilder-Smith, *Creation of Life*, 69, 86.
77 A. Wilder-Smith, 218.
78 A. Wilder-Smith, 234.
79 This lacuna in mid-twentieth-century creationist writing is discussed more in chapter 2. ID advocates have, for their part, read Wilder-Smith as an exception rather than a rule in creationist thinking, since he focused on the design argument more than young-Earth creationists usually did. Some major ID figures have acknowledged him as an influence. William Dembski wrote, "It's worth noting that the effort to make the design of natural systems scientifically tractable has at best been a peripheral concern of young-Earth creationists historically. There have been exceptions, like A. E. Wilder-Smith, who sought to identify the information in biological systems and connect it with a designer/creator. But the principal texts of the Institute for Creation Research, for instance, typically took a very different line from trying to make design a program of scientific research. Instead of admitting that Darwinian theory properly belonged to science and then trying to formulate design as a replacement theory, young-Earth creationists typically claimed that neither Darwinism nor design could properly be regarded as scientific (after all, so the argument went, no one was there to observe what either natural selection or a designer did in natural history)." Dembski, "Intelligent Design Coming Clean." In addition, Phillip Johnson called Wilder-Smith a "prophet" who "anticipated intelligent design." See Kushiner, "Measure of Design," 17. It is important to note, though, that just because creationists did not center design does not mean that they avoided it. Harry Rimmer, for instance, frequently

used design in his polemics against Darwin. See Numbers, *Creationists*, 65. That said, design was usually not the focal point of young-Earth creationist attacks.

80 Pearcey, "Up from Materialism," 7.

81 Kenyon, foreword to *What Is Creation Science?*, 11.

82 Morris and Parker, *What Is Creation Science?*, 5. Duane Gish, who held a PhD in biochemistry from Cal, also made this argument frequently.

83 Morris and Parker, 52–54.

84 Morris and Parker, 112.

85 Morris and Parker, 39, 77.

86 Eldredge, *Triumph of Evolution*, 12.

87 *McLean, v. Arkansas*, 529 F. Supp. 1255, 1272 (1982).

88 See Ruse, *But Is It Science?*, for a detailed explanation, from Ruse himself, on why he determined that the demarcation problem (what classifies or does not classify as science) was the best strategy to defeat creationism. Ruse included in the book a response from the philosopher of science Larry Laudan, who objected to this strategy because he felt that creationism should be thrown out not because of its failure to satisfy abstract definitions of science in philosophy but because it was empirically wrong and could be shown to be so. Laudan worried that a failure to dismiss creationism as scientifically vacuous would open the door for future anti-evolutionary movements.

89 Edward Larson, *Trial and Error*, 147.

90 Edward Larson, 160.

91 Elsberry and Matzke, "Collapse of Intelligent Design," 76–78.

92 Weatherly, "Creationists Lose in Arkansas," 29.

93 Lewin, "Creationism on the Defensive in Arkansas," 34.

94 Frair, preface to *Creation and the Courts*, 12.

95 *McLean*, 529 F. Supp. at 1270.

96 Geisler, *Creation and the Courts*, 24.

97 *Edwards v. Aguillard*, 482 U.S. 578 (1987).

98 Kenyon, foreword to *Mystery of Life's Origin*, viii.

99 Numbers, *Creationists*, 374; Woodward, *Doubts About Darwin*, 87.

100 Bradley, Olsen, and Thaxton, *Mystery of Life's Origin: Reassessing Current Theories*, 188–214.

101 Bradley, Olsen, and Thaxton, 8.

102 Meyer, *Signature in the Cell*, 24–34.

103 Meyer, 4.

104 Woodward, *Doubts About Darwin*, 86.

105 Gish, *Speculations and Experiments*; Wysong, *Creation-Evolution Controversy*.

106 Geisler, *Origin Science*, 8, 139.

107 Geisler, 145–146.

108 Geisler, 74.

109 Geisler, 153.

110 Geisler, 14–15.

111 Bird, *Origin of Species Revisited*, 1:46–69, 295–392, 393–474.
112 Denton, *Evolution*.
113 Johnson, "Book Review," 428.
114 Johnson, 430.
115 *Of Pandas and People*'s origins were drawn out and exposed most clearly in the proceedings and summary of the *Dover* trial in 2005. See Judge Jones's opinion, *Kitzmiller v. Dover Area School Dist.*, 400 F. Supp. 2d 707. See also chapter 4 of this book.

CHAPTER 2. DESIGN

1 "Berkeley's Radical." Some of these "genesis of ID" narratives differ in the details. Thomas Woodward described the story as beginning when Johnson saw Denton's and Dawkins's works next to each other in the bookstore and does not mention Asimov. See Woodward, *Doubts About Darwin*, 69–70. Again, slightly differently, Stephen Meyer described Johnson's doubts beginning when he went to the British Natural History Museum in London. See Meyer, "Your Witness, Mr. Johnson," 33–34.
2 Woodward, *Doubts About Darwin*, 74–79.
3 Johnson, "Appendix 2," 218–219.
4 Johnson, 224.
5 Woodward, *Doubts About Darwin*, 81.
6 Dembski, preface to *Darwin's Nemesis*, 13.
7 Reynolds, "Introduction," 25. See Numbers, *Creationists*, 390.
8 Luskin, "Why Phillip Johnson Matters."
9 Reynolds, "Introduction," 28–30.
10 Dembski, preface to *Darwin's Nemesis*, 14–15.
11 Behe, *Darwin Devolves*, 9.
12 Waldrop, *Dream Machine*, 18.
13 Waldrop, 224.
14 Johnson, *Darwin on Trial*, 21 (emphasis in original).
15 Johnson, 144–145.
16 Johnson, "Evolution as Dogma."
17 Johnson, "Response to Gould."
18 Dawkins, *Blind Watchmaker*, 1.
19 Dawkins, *River Out of Eden*, 133.
20 Johnson, *Reason in the Balance*, 14–15.
21 Dembski, introduction to *Mere Creation*, 27.
22 Lewontin, "Billions and Billions of Demons."
23 Dembski, "Dealing with the Backlash," 100.
24 Behe, foreword to *Darwin on Trial*, 12.
25 Berlinski, "Deniable Darwin," 62.
26 Keller, "Wedge of Truth Visits the Laboratory," 156.
27 Haught, "Review."

28 Pennock, *Tower of Babel*, 191.
29 Ward, *Big Questions in Science and Religion*, 71.
30 Woodward, *Doubts About Darwin*, 149.
31 Dembski, *Intelligent Design*, 119.
32 Numbers, *Creationists*, 274–279.
33 Johnson, *Darwin on Trial*, 179–189. For an overview, see Sonleitner, "What Did Karl Popper Really Say About Evolution?"
34 Johnson, "Afterword," 450–451.
35 Feyerabend, "How to Be a Good Empiricist," 922–949.
36 Kuhn, *Structure of Scientific Revolutions*.
37 For an overview of Kuhn and a summary of criticism concerning his thesis, see Godfrey-Smith, *Theory and Reality*, chapters 5 and 6.
38 Marsden, *Fundamentalism and American Culture*, 214.
39 Denton, *Evolution*, 348, 351, 356.
40 Johnson, *Darwin on Trial*, 149–152.
41 Witt, "March for Science?"
42 Luskin, "Jonathan Wells Talks Scientific Revolutions."
43 DeWolf et al., *Traipsing into Evolution*, 46–47.
44 Luskin, "Jonathan Wells Talks Scientific Revolutions."
45 Wells, *Politically Incorrect Guide to Darwinism*, 196–197.
46 Godfrey-Smith, *Theory and Reality*, 95–96.
47 Pennock, *Tower of Babel*, 210.
48 Wells, "Is Darwinism a Theory in Crisis?," 414.
49 Wells, *Zombie Science*, 185.
50 Chapman, postscript to *Mere Creation*, 459.
51 Wells, "Unseating Naturalism," 68.
52 Dembski, *Intelligent Design*, 119.
53 Denton, *Evolution*, 356.
54 Wells, "Unseating Naturalism," 68.
55 Johnson, *Reason in the Balance*, 12.
56 Johnson, *Darwin on Trial*, 37, 148, 238.
57 Behe, "From Muttering to Mayhem."
58 Behe, *Darwin's Black Box*, xii.
59 Darwin, *Origin of Species*, 189.
60 Behe, *Darwin's Black Box*, 3.
61 Behe, 39.
62 Behe, 186.
63 Twain, *Roughing It*, 41.
64 Coyne, "God in the Details," 227.
65 K. Miller, "*Darwin's Black Box* Reviewed."
66 K. Miller, 38.
67 Behe, *Edge of Evolution*, 103–104.
68 Behe, *Mousetrap for Darwin*, 63.

69 Dembski, *Design Revolution*, 294–298. Here Dembski summarized and defended Behe's concept of irreducible complexity. The same point about the nonexistence of testable or known indirect pathways for the evolution of complex biochemical systems can be found in Behe, *Darwin's Black Box*, 165–186.

70 Behe, *Darwin's Black Box*, 193, 208, 233.

71 Geisler, "Review of Michael Behe's *The Edge of Evolution*."

72 Coyne, "God in the Details," 227.

73 Geisler, "Review of Michael Behe's *The Edge of Evolution*."

74 Dembski, "Intelligent Design Coming Clean."

75 Dembski, *Being as Communion*, 92–93.

76 The details of Dembski's conversions are obscure, and he did not reference them much. When he appeared in news articles around the turn of the millennium, he was often referred to as an Orthodox layman, but this was not a term he self-applied. See Carnes, "Intelligent Design." See also a short overview in Numbers, *Creationists*, 384. It is probable that Dembski's interest in Orthodoxy was behind his editing a collection of patristic writings on the doctrine of creation. See Dembski et al., *Patristic Understanding of Creation*. In 2019, Dembski also recommended the Eastern Orthodox spiritual handbook the *Philokalia* to people interested in learning more about Christian theology. See Barham, "William Dembski Interview."

77 Johnson, "Final Word," 317; Forrest and Gross, *Creationism's Trojan Horse*, 143.

78 Dembski, *Design Inference*, 36.

79 Dembski, 41.

80 Stoppard, *Rosencrantz and Guildenstern Are Dead*, 11–15.

81 Dembski, *Design Inference*, 55–62.

82 Dembski, 62, 66.

83 Dembski, *Intelligent Design*, 128.

84 Dembski, 128.

85 Sample, "Stephen Hawking Launches $100m Search."

86 Dembski, *Design Revolution*, 270–273.

87 Dembski, 276–277.

88 Dembski, 317.

89 Dembski, *No Free Lunch*.

90 Paley, *Natural Theology*, 8.

91 Paley, 16.

92 Ruse, "Argument from Design," 38.

93 Grant, "Intentional Deception."

94 Pennock, "DNA by Design?," 140–141.

95 Numbers, *Darwinism Comes to America*, 18.

96 Woodward, *Doubts About Darwin*, 187.

97 Denton, *Evolution*, 341.

98 Behe, *Darwin Devolves*, 229.

99 Behe, *Darwin's Black Box*, 211–216.

100 Woodward, *Doubts About Darwin*, 202.
101 Axe, "Is Our Intuition of Design in Nature Correct?," 152.
102 Meyer, *Return of the God Hypothesis*, 261.
103 Luskin, "No, Intelligent Design Doesn't Reason by Analogy."
104 Behe, *Mousetrap for Darwin*, 37 (emphasis in original).
105 Lang and Rice, "Evolution Unscathed"; Behe, *Mousetrap for Darwin*, 445–446.
106 Behe, *Mousetrap for Darwin*, 19.
107 Behe, "How Does Irreducible Complexity Challenge Darwinism?," 301–302.
108 Denton, "Anti-Darwinian Intellectual Journey," 170.
109 Dembski, *Being as Communion*, 61–62.
110 Hume, *Dialogues and Natural History of Religion*, 78–82.
111 Mumford, *Pentagon of Power*, 87, 91.
112 Mumford, 391–392.
113 Dembski, *Being as Communion*, 114.
114 Saunders, *Divine Action and Modern Science*, 8.
115 Saunders, 211.
116 Saunders, 214–215.
117 Shapiro, "Myth 8," 68.
118 See also Manning, introduction to *Oxford Handbook of Natural Theology*. Manning writes that the usual narrative of Paley's decline and Darwin's rebuttal is an ungrounded myth and should be rejected, "no matter how widely accepted it is" (2).
119 Topham, "Biology in the Service of Natural Theology," 91.
120 Numbers, *Creation by Natural Law*. Darwin held to the "creation by law" view of nature for a time as well. See Desmond and Moore, *Darwin*, 293.
121 Bowler, "Darwinism and the Argument from Design."
122 LeMahieu, *Mind of William Paley*, x.
123 Shapiro, "Myth 8," 69.
124 Dembski, "Is Intelligent Design a Form of Natural Theology?"
125 Dembski.
126 Charles Darwin to Asa Gray, May 22, 1860.
127 Gilbert, "Aerodynamics of Flying Carpets," 51.
128 Dembski, *Intelligent Design*, 261–262.
129 Behe, "Bullet Points for Jerry Coyne."
130 Coyne, *Why Evolution Is True*, 82–84.
131 Dawkins, *Greatest Show on Earth*, 384.
132 Egnor, "Welcome, Jerry Coyne."
133 Kenyon and Davis, *Of Pandas and People*, 120–121.
134 Gould, *Panda's Thumb*, 19–26.
135 Axe, *Undeniable*, 78.
136 Eberlin, *Foresight*, 120–21.
137 Coyne, "ID Craziness."
138 Dembski, *Intelligent Design*, 206.

139 Forrest and Gross, *Creationism's Trojan Horse*, 261.
140 Eldredge, *Triumph of Evolution*, 135.
141 Ratzsch, "How Not to Critique Intelligent Design Theory," 192–193. This changed slightly in Numbers's second edition of *The Creationists*, from 2006, which included a chapter on ID and a short discussion of William Paley's influence.
142 A. E. Wilder-Smith is another example, covered in the previous section. As noted earlier, Dembski drew attention to Wilder-Smith in an interview where he acknowledged the influence of a few creationists, Wilder-Smith being one of them.
143 Numbers, *Creationists*, 325–326.
144 Morris, "Intelligent Design and/or Scientific Creationism."
145 Morris, *History of Modern Creationism*, 124.
146 Geisler, "Review of Michael Behe's *The Edge of Evolution*."
147 Ham, "Intelligent Design Is Not Enough."
148 Dembski, *Intelligent Design*, 107.
149 Ayala, "Design Without Designer," 69.

CHAPTER 3. POLITICS

1 LeMahieu, *Mind of William Paley*, 132–133.
2 Desmond, *Politics of Evolution*, 110–111.
3 Cobbett, *Selections from Cobbett's Political Works*, 131. For more on William Cobbett, see Howell, "William Cobbett and American Society." See also Ingrams, *Life and Adventures of William Cobbett*.
4 Reynolds, "Introduction," 26.
5 Johnson, *Reason in the Balance*, 13, 17.
6 Johnson, 48.
7 Discovery Institute, "Major Grants Help Establish Center."
8 Linker, *Theocons*.
9 See Secord, *Victorian Sensation*.
10 Adam Sedgwick to Charles Lyell, April 9, 1845, 85. See also Ruse, *But Is It Science?*, 58–61; and Desmond and Moore, *Darwin*, 322.
11 Brooke, "Richard Owen, William Whewell, and the *Vestiges*." See this article for a reproduction of the letters between Owen and Whewell.
12 Sedgwick, "*Vestiges of the Natural History of Creation*," 3.
13 Chambers, *Vestiges of the Natural History of Creation*, 26, 324.
14 Secord, *Victorian Sensation*; Desmond and Moore, *Darwin*, 320–325.
15 Kirk, *Conservative Mind*, 8.
16 Scruton, *Conservatism: Invitation to the Great Tradition*, 144.
17 Koons, "Check Is in the Mail," 20.
18 Birzer, *Russell Kirk*, 137.
19 Nash, *Conservative Intellectual Movement*, 52.
20 Weaver, *Visions of Order*, 137.
21 Kirk, foreword to *Visions of Order*, ix.
22 Weaver, *Visions of Order*, 138–139 (emphasis in original).

23 Weaver, 143.
24 Koons, "Check Is in the Mail," 12.
25 Klinghoffer, "No Real Conflict When One Side Gives Up."
26 Voegelin, *New Order and Last Orientation*, 184–185 (also available online at http://voegelinview.com).
27 Nisbet, *Prejudices*, 79.
28 Buckley, *God and Man at Yale*, 181.
29 Firinglinevideos, "Firing Line Debate."
30 Buckley, "So Help Us Darwin."
31 Vaïsse, *Neoconservatism*.
32 Himmelfarb, *Darwin and the Darwinian Revolution*, 312, 316, 331, 338, 442–443 (emphasis in original).
33 Kristol, "Room for Darwin and the Bible."
34 Futuyma, "Evolution."
35 Berlinski, "Deniable Darwin," 56.
36 Berlinski, 62.
37 For a collection of letters to the editor and Berlinski's reply, see "Denying Darwin: David Berlinski & Critics," in Klinghoffer, *Deniable Darwin and Other Essays*, 65–141.
38 "Denying Darwin," 109–110.
39 Berlinski, *Devil's Delusion*, xiii.
40 Berlinski, "Has Darwin Met His Match?" For the letters to the editor, see "Darwinism and Intelligent Design: David Berlinski & Critics," in Klinghoffer, *Deniable Darwin and Other Essays*, 313–366.
41 Bailey, "Origin of the Specious."
42 Fawcett, *Conservatism*, 19.
43 Himmelfarb, *Darwin and the Darwinian Revolution*, 31, 410.
44 Kristol, *Neoconservatism*, 4, 134.
45 For the full account, see Linker, *Theocons*.
46 "Editorial: Putting First Things First."
47 Johnson, "Evolution as Dogma."
48 Johnson, *Reason in the Balance*, 12–13.
49 Fawcett, *Conservatism*, 47.
50 Gilder, "Evolution and Me."
51 Anderson, "Evolution of a Think Tank."
52 Linker, *Theocons*, 186.
53 Neuhaus, "Bashing Darwin, Becoming Catholic."
54 Oakes, "Wedge of Truth."
55 Arnhart, in "Conservatives, Darwin, & Design."
56 Behe, in "Conservatives, Darwin, & Design."
57 West, *Darwinian Conservatives*. See also West, *Darwin Day in America*.
58 West, *Darwin Day in America*, 29, 116–117.
59 Will, *Conservative Sensibility*, 216–298.

60 Hayek, *Fatal Conceit*, 76–77.
61 Scruton, *Conservatism*, 109.
62 Will, *Conservative Sensibility*, 244, 504.
63 West, *Darwin Day in America*, 115–118.
64 Wells, *Politically Incorrect Guide to Darwinism*, 166–167.
65 Numbers, *Creationists*, 274. See also Weinberg, "Ye Shall Know Them by Their Fruits."
66 Morris, *Long War Against God*, 58–59.
67 Hayek, *Constitution of Liberty*.
68 Toumey, *God's Own Scientists*, 6, 77, 98–99.
69 Nash, *Conservative Intellectual Movement*, 361.
70 J. Evans, *Morals Not Knowledge*, 77.
71 J. Evans, 79.
72 J. Evans, 21.
73 Nash, *Conservative Intellectual Movement*, 368.
74 Fawcett, *Conservatism*, 368–370.
75 Buchanan, "Darwin's Pyrrhic Victory."
76 Buchanan, "Making a Monkey out of Darwin."
77 Lemonick, "Dumping on Darwin."
78 Buchanan, "Darwin's Pyrrhic Victory."
79 Belluck, "Board for Kansas Deletes Evolution." The thrust to delete evolution was more populist creationist than it was ID, for the board also removed references to the Big Bang theory, which most ID advocates actually support.
80 Frank, *What's the Matter with Kansas?*, 240.
81 Frank, 213–214.
82 Marsden, *Fundamentalism and American Culture*, 246.
83 Dembski, *Design Revolution*, 306–309.
84 Ratliffe, "Crusade Against Evolution."
85 *Selman v. Cobb County*, 449 F.3d 1320, 1324 (11th Cir. 2006).
86 For a brief synthesis of these events, see Humes, *Monkey Girl*, 146–160. Humes's book is a chronicle of the 2005 *Dover* trial but covers the Kansas events as a harbinger of the court case.
87 147 Cong. Rec. S6148 (daily ed. June 13, 2001).
88 Discovery Institute, "Key Resources."
89 "Evangelical Scientists Refute Gravity."
90 147 Cong. Rec. S6152 (daily ed. June 13, 2001).
91 Wald, "Intelligent Design Meets Congressional Designers."
92 Branch, "Farewell to the Santorum Amendment?"
93 *Kitzmiller v. Dover*, 400 F. Supp. 2d 707, 708 (M.D. Pa. 2005).
94 Colson, foreword to *The Design Revolution*, 16.

CHAPTER 4. BACKLASH

1 Quoted in Coyne, "Jason Rosenhouse Pronounces Intelligent Design Dead."
2 Couzin-Frankel, "Bush Backs Teaching Intelligent Design."

3 There was considerable controversy about Flew's conversion, not least because his subsequent book *There Is a God* was cowritten by Roy Abraham Varghese. The journalist Mark Oppenheimer accused Varghese of authoring the entire book and alleged that Flew was in a state of mental decline. See Oppenheimer, "Turning of an Atheist." Flew responded by criticizing Oppenheimer. In 2006, he appeared at an ID conference at Biola University, where he was presented with the inaugural Phillip E. Johnson Award for Liberty and Truth. See Witt, "Biola Honors Flew."
4 Humes, *Monkey Girl*, 225–227.
5 This struggle was evident in Kansas, too. The first time evolution was struck from the school board, it was not the only theory to fall by the wayside—the Big Bang theory was also removed from the curriculum, an irony given how frequently ID advocates would draw on the Big Bang during *Dover*.
6 Woodward, *Doubts About Darwin*, 14.
7 Behe, "Teach Evolution."
8 Matzke, "Design on Trial in Dover."
9 Humes, *Monkey Girl*, 44.
10 Slack, *Battle over the Meaning of Everything*, 11.
11 Humes, *Monkey Girl*, 76–78.
12 Matzke, "Design on Trial in Dover."
13 Banerjee, "School Board Sued"; Matzke, "Design on Trial in Dover."
14 Humes, *Monkey Girl*, 192.
15 Matzke, "Design on Trial in Dover."
16 Humes, *Monkey Girl*, 217.
17 Hunter, *Culture Wars*, 35–39, 47–48.
18 M. Chapman, *40 Days and 40 Nights*, 21–22, 270.
19 M. Chapman, 144.
20 Burress, "Teaching Evolution as Theory Not Fact."
21 Discovery Institute, "Discovery Calls Dover Evolution Policy Misguided."
22 Humes, *Monkey Girl*, 239–241.
23 Slack, *Battle over the Meaning of Everything*, 85–93.
24 Slack, 73.
25 Humes, *Monkey Girl*, 76.
26 Goodstein, "Evolution Trial in Hands of Willing Judge."
27 Dembski, "Life After Dover."
28 Talbot, "Darwin in the Dock."
29 Humes, *Monkey Girl*, 337–338.
30 "Morning Session—Transcript of the Proceedings of Bench Trial," *Kitzmiller v. Dover*, September 26, 2005, 9–10, 14.
31 "Morning Session," 27.
32 "Morning Session," 58.
33 "Morning Session—Transcript of the Proceedings of Bench Trial," *Kitzmiller v. Dover*, September 28, 2005, 10–12.

34 "Defendants' Brief in Support of Motion *In Limine* to Exclude the Testimony of Barbara Forrest, Ph.D.," *Kitzmiller, v. Dover*, September 6, 2005.
35 Forrest's testimony was covered extensively in most retrospectives of the trial. See Humes, *Monkey Girl*, 284–292; Slack, *Battle over the Meaning of Everything*, 98–108; and Chapman, *40 Days and 40 Nights*, 138–149. For historical assessments, see Numbers, *Creationists*, 391–394. For an assessment from Forrest herself, see Forrest and Gross, *Creationism's Trojan Horse*, 317–338.
36 "Morning Session," September 28, 2005, 51.
37 "Afternoon Session—Transcript of the Proceedings of Bench Trial," *Kitzmiller v. Dover*, September 30, 2005, 7.
38 "Afternoon Session—Transcript of the Proceedings of Bench Trial," *Kitzmiller v. Dover*, October 14, 2005, 47–48.
39 "Morning Session," September 26, 2005, 63–64.
40 "Afternoon Session," September 30, 2005, 30.
41 "Afternoon Session," October 14, 2005, 41.
42 "Morning Session," September 27, 2005, 61–69.
43 Chapman, *40 Days and 40 Nights*, 51.
44 Humes, *Monkey Girl*, 237.
45 "Afternoon Session," September 30, 2005, 49–53.
46 Humes, *Monkey Girl*, 272.
47 Chapman, *40 Days and 40 Nights*, 112.
48 Humes, *Monkey Girl*, 297.
49 Lebo, *Devil in Dover*, 151.
50 Humes, *Monkey Girl*, 299.
51 "Afternoon Session—Transcript of the Proceedings of Bench Trial," *Kitzmiller v. Dover*, October 18, 2005, 39.
52 "Afternoon Session," 86; see also Humes, *Monkey Girl*, 302–305.
53 "Afternoon Session—Transcript of the Proceedings of Bench Trial," *Kitzmiller v. Dover*, October 19, 2005, 14–16.
54 Slack, *Battle over the Meaning of Everything*, 143.
55 Humes, *Monkey Girl*, 307.
56 "Morning Session—Transcript of the Proceedings of Bench Trial," *Kitzmiller v. Dover*, October 24, 2005, 129.
57 "Afternoon Session—Transcript of the Proceedings of Bench Trial," *Kitzmiller v. Dover*, October 24, 2005, 42.
58 "Afternoon Session—Transcript of the Proceedings of Bench Trial," *Kitzmiller v. Dover*, November 3, 2005, 52.
59 Humes, *Monkey Girl*, 307.
60 "Morning Session—Transcript of the Proceedings of Bench Trial," *Kitzmiller v. Dover*, October 27, 2005, 80, 90–92.
61 Humes, *Monkey Girl*, 309; Lebo, *Devil in Dover*, 84.
62 "Morning Session," October 27, 2005, 96–97.

63 "Afternoon Session—Transcript of the Proceedings of Bench Trial," *Kitzmiller v. Dover*, October 27, 2005, 46–47.
64 "Afternoon Session," 6.
65 Humes, *Monkey Girl*, 315.
66 Humes, 331.
67 *Kitzmiller v. Dover*, 400 F. Supp. 2d 707, 765 (M.D. Pa. 2005).
68 *Kitzmiller*, 400 F. Supp. 2d at 756.
69 *Kitzmiller*, 400 F. Supp. 2d at 739.
70 *Kitzmiller*, 400 F. Supp. 2d at 765.
71 *Kitzmiller*, 400 F. Supp. 2d at 722–723.
72 DeWolf et al., *Traipsing into Evolution*, 9.
73 Lebo, *Devil in Dover*, 195.
74 *Kitzmiller*, 400 F. Supp. 2d at 765.
75 Powell, "Judge Rules Against 'Intelligent Design.'"
76 *Kitzmiller*, 400 F. Supp. 2d at 720.
77 Dawkins, *God Delusion*, 57.
78 Simonyi, "Manifesto for the Simonyi Professorship."
79 Johnson, "Response to Gould."
80 Dawkins, *Brief Candle in the Dark*, 173.
81 Bullivant, "New Atheism and Sociology," 110.
82 Ali, "Why I Am Now a Christian."
83 Wolf, "Church of the Non-Believers." Irritated by the puffed-up tone of the New Atheism and the self-congratulatory name of "The Four Horsemen," the Christian philosopher Alvin Plantinga wondered whether a better title, since Sam Harris was at the time a lowly graduate student, might be "The Three Bears of Atheism." See Plantinga, "Dawkins Confusion."
84 See Amarasingam, *Religion and the New Atheism*; Cotter et al., *New Atheism*; McAnulla et al., *Politics of New Atheism*; Numbers and Hardin, "New Atheists."
85 Dawkins, "Religion's Misguided Missiles."
86 Dawkins, "Time to Stand Up."
87 Harris, *End of Faith*, 109–110; Ali, *Infidel*.
88 Dawkins, *God Delusion*, 345–346.
89 Stenger, *New Atheism*, 241.
90 Dawkins, *God Delusion*, 103, 138.
91 Dawkins, *Greatest Show on Earth*, 4, 7–8.
92 Hitchens, *God Is Not Great*, 54, 82, 91.
93 Harris, *Letter to a Christian Nation*, 72.
94 Dennett, *Breaking the Spell*, 61.
95 Dembski, introduction to *Mere Creation*, 15.
96 Johnson, *Reason in the Balance*, 208–209.
97 Meyer, "Return of the God Hypothesis," 27.
98 Dawkins, *God Delusion*, 52, 189.
99 Stenger, *God—The Failed Hypothesis*, 11.

100 Stenger, *New Atheism*, 102.
101 Johnson and Reynolds, *Against All Gods*, 8–9.
102 Meyer, *Return of the God Hypothesis*, 6.
103 Meyer, 236.
104 Dawkins, *God Delusion*, 93–94; Bunting, "Why the Intelligent Design Lobby Thanks God."
105 Clayton, *Religion and Science*, 15–23; Numbers and Hardin, "New Atheists," 225–226.
106 Barbour, *Religion and Science*, 81, 84.
107 Cunningham, *Darwin's Pious Idea*, 275.
108 J. Evans, *Morals Not Knowledge*, 24.
109 Hart, "Believe It or Not."
110 Alexander, "New Atheism"; A. Nichols, "New Atheism's Idiot Heirs."
111 Myers, "Train Wreck That Was the New Atheism."
112 Pigliucci, *How to Be a Stoic*, 89.
113 Poole, "Four Horsemen Review."
114 Came, "Richard Dawkins's Refusal to Debate"; Elmhirst, "Is Richard Dawkins Destroying His Reputation?"
115 Lundmark and LeDrew, "Unorganized Atheism and the Secular Movement."
116 Feser, *Five Proofs of the Existence of God*, 287–288.
117 Behe, *Edge of Evolution*, 229.
118 Dembski, introduction to *Mere Creation*, 20.
119 Denton, *Evolution*, 66.
120 Johnson, *Wedge of Truth*, 100.
121 Quoted in Forrest and Gross, *Creationism's Trojan Horse*, 296.
122 Johnson, *Wedge of Truth*, 95.
123 Forrest and Gross, *Creationism's Trojan Horse*, 288.
124 Dembski, introduction to *Mere Creation*, 21.
125 Meyer, *Return of the God Hypothesis*, 279.
126 Meyer, 219.
127 Meyer, 241.
128 Behe, *Mousetrap for Darwin*, 66–67.
129 Meyer, *Return of the God Hypothesis*, 295.
130 Johnson, *Wedge of Truth*, 89.
131 K. Miller, *Finding Darwin's God*; Haught, *God After Darwin*; Humes, *Monkey Girl*, 192.
132 *Kitzmiller v. Dover*, 400 F. Supp. 2d 707, 765 (M.D. Pa. 2005).
133 F. Collins, *Language of God*, 195.
134 F. Collins, 203.
135 Sullivan, "Helping Christians Reconcile God with Science."
136 Some of the New Atheists expressed irritation and even anger at Collins's appointment. In a *New York Times* piece, Sam Harris wrote, "Francis Collins is an accomplished scientist and a man who is sincere in his beliefs. And that is precisely what makes me so uncomfortable about his nomination." Harris, "Science Is in the Details."

137 See John Templeton Foundation, "BioLogos Public Engagement"; John Templeton Foundation, "Engaging the Church and the World."
138 Goodstein, "Intelligent Design Might Be Meeting Its Maker."
139 Rosenhouse, "Templeton Foundation and Santorum Losing Faith in ID?" See also "Wall Street Journal Wrong on Templeton Foundation and ID."
140 I owe this observation to Francis Beckwith, who pointed it out to me. See William Dembski's website for the list of grants he has received. Dembski, "Life Work of Bill Dembski."
141 Winfield, "Vatican-Backed Conference Snubs Creationism."
142 Schönborn, "Finding Design in Nature."
143 Currid, "Theistic Evolution Is Incompatible," 842–843.
144 Rana, "What Is the Biblical and Scientific Case?," 99.
145 Inter Varsity Press, "BioLogos Books on Science and Christianity."
146 Van Biema, "God vs. Science."
147 Ecklund and Scheitle, *Religion vs. Science*, 43, 49.
148 Ecklund and Scheitle, "Influence of Science Popularizers."
149 Meyer, "Scientific and Philosophical Introduction," 43, 46.
150 Meyer, 54.
151 Grudem, "Biblical and Theological Introduction," 62.
152 Grudem, "Theistic Evolution Undermines Twelve Creation Events."
153 Fuller, foreword to *Theistic Evolution*, 30–31.
154 Currid, "Theistic Evolution Is Incompatible," 860, 862.
155 Reeves, "Bringing Home the Bacon," 722.
156 Currid, "Theistic Evolution Is Incompatible," 878.
157 J. Evans, *Morals Not Knowledge*, 89.
158 Beckwith, *Never Doubt Thomas*, 60.
159 Beckwith, 62.
160 Beckwith, 63.
161 Beckwith, 76.
162 Beckwith, 65–66.
163 Beckwith, 79.
164 Meyer, "Difference It Doesn't Make," 217–219.
165 Meyer, "Scientific and Philosophical Introduction," 43.
166 Meyer, 44–45.
167 Meyer and Nelson, "Should Theistic Evolution Depend on Methodological Naturalism?," 565.
168 Geisler, *Origin Science*, 144.
169 C. Collins, "How to Think about God's Action in the World," 672.
170 C. Collins, 680 (emphasis in original).
171 Grudem, "Theistic Evolution Undermines Twelve Creation Events," 813 (emphasis in original).
172 Grudem, 814 (emphasis in original).
173 Hart, *Experience of God*, 35–38.

174 Hart, 37–38.
175 Koons and Gage, "St. Thomas Aquinas on Intelligent Design," 91–92.
176 Koons and Gage, 81.
177 Koons and Gage, 82, 91.
178 Koons and Gage, 91.
179 Koons and Gage, 82, 85.
180 Beckwith, *Never Doubt Thomas*, 84–85.
181 Dembski, *Being as Communion*, 25.
182 Dembski, 58–59.
183 Dembski, 58–59n27.
184 Feser, *Scholastic Metaphysics*, 25.
185 Feser, 50.
186 Feser, *Aristotle's Revenge*, 43–44, 48.
187 Feser, 432–433.
188 Feser, 434–435.
189 Feser, 436.
190 Feser, 441–442.
191 Dembski, "Naturalism's Argument from Invincible Ignorance."
192 Van Till, "SOBIG: A Symposium on Belief in God," 34.
193 Forrest and Gross, *Creationism's Trojan Horse*, 338.

CHAPTER 5. AFTERMATH

1 Rosenhouse, "Twenty Years After Darwin on Trial." See also Rosenhouse, *Among the Creationists*.
2 Padian, "Lessons from the 'Intelligent Design' Trial," 93.
3 P. Wallace, "Intelligent Design Is Dead."
4 Klinghoffer, "New Rallying Cry."
5 Luskin, "How Bright Is the Future of Intelligent Design?"; Luskin, "It's Time for Some Folks to Get Over Dover."
6 Axe, "Extreme Functional Sensitivity."
7 Venema, "Intuitions, Proteins, and Evangelicals."
8 Axe, *Undeniable*, 51.
9 National Center for Science Education, "Wedge Document."
10 Biever, "God Lab," 11.
11 Bailey, "BioLogic Institute = Tobacco Institute."
12 Axe, Dixon, and Gauger, "Good Science Will Come."
13 Branch, "Latest 'Intelligent Design' Journal."
14 "Editorial Policies."
15 Richards, "*BIO-Complexity*."
16 "Editorial Policies."
17 Shallit, "Sterility of Intelligent Design."
18 Shallit, "Groundless Annual Ritual."
19 Branch, "Latest 'Intelligent Design' Journal."

20 Torley, "*Undeniable* Packs a Strong Punch."
21 Meyer, *Signature in the Cell*, 24–31.
22 Meyer, "Danger: Indoctrination."
23 Discovery Institute, "Major Grants Help Establish Center for Renewal of Science and Culture."
24 Humes, *Monkey Girl*, 170.
25 Meyer, *Signature in the Cell*, 2–7.
26 Meyer, 13.
27 Meyer, 137–138.
28 Meyer, 85–323.
29 Meyer, 150–153.
30 Meyer, 347–348.
31 Meyer, 376–379.
32 Quoted in West, "Signature in the Cell Named One of Top Books."
33 "TLS Letters 09/12/09."
34 Nagel, *Mind and Cosmos*, 10.
35 Rosenhouse, "Twenty Years After Darwin on Trial"; Hoppe, "Two Analyses."
36 Klinghoffer, *Signature of Controversy*.
37 Meyer, *Darwin's Doubt*, vii–ix.
38 Meyer, 185–208, 240–253.
39 Meyer, 359, 389, 392.
40 Klinghoffer, "Introduction: Intelligent Design's Original Edition," 13.
41 Behe, *Edge of Evolution*, 84.
42 Behe, 94–95.
43 Behe, 210–216.
44 Dawkins, "Inferior Design."
45 Behe, *Darwin Devolves*, 229–238.
46 Behe, 37–38.
47 Coyne, "Intelligent Design Gets Even Dumber."
48 Lents, "Behe's Last Stand."
49 Wells, *Icons of Evolution*, 157.
50 Wells, 225, 241.
51 Coyne, "Creationism by Stealth," 745–746.
52 Wells, *Zombie Science*, 59, 65.
53 Wells, 177.
54 Wells, 178, 187.
55 Klinghoffer, "Introduction: Intelligent Design's Original Edition," 13.
56 Marks and West, foreword to *The Mystery of Life's Origin*, 8.
57 Thaxton, "Appendix 2: 1997 Update," 311–314.
58 Tour, "We're Still Clueless"; B. Miller, "Thermodynamic Challenges"; Gonzalez, "What Astrobiology Teaches."
59 Wells, "Textbooks Still Misrepresent the Origin of Life," 407–410.
60 Wells, 394.

61 Meyer, "Evidence of Intelligent Design in the Origin of Life," 439.
62 Tour, "We're Still Clueless," 327, 335.
63 Frankowski, *Expelled.*
64 Frankowski.
65 Chapman, "An Intelligent Discussion about Life."
66 Frankowski, *Expelled.*
67 Martin, "Professors Debate Legitimacy of Polanyi."
68 Schmeltekopf, *Baylor at the Crossroads*, 77.
69 Baylor University External Review Committee, "External Review Committee Report."
70 Dembski, "Polanyi Center Press Release."
71 Brumley, "Dembski Relieved of Duties."
72 Heeren, "Lynching of Bill Dembski."
73 Wells, *Politically Incorrect Guide*, 91.
74 DeWolf et al., *Traipsing into Evolution*, 51.
75 Beckwith, "Review—*The Creationists*," 738.
76 Woodward, *Doubts About Darwin*, 180.
77 *Kitzmiller v. Dover*, 400 F. Supp. 2d 707, 735 (M.D. Pa. 2005).
78 Wells, *Politically Incorrect Guide*, 104.
79 Meyer, "Origin of Biological Information," 234.
80 Biological Society of Washington, "Statement from the Council."
81 Sternberg, "Procedures for the Publication of the Meyer Paper."
82 Discovery Institute, "U.S. Congressional Committee Report."
83 Klinghoffer, "Branding of a Heretic."
84 Meyer, *Darwin's Doubt*, 384–385, 402.
85 Frankowski, *Expelled.*
86 Frankowski.
87 Bailey, "Flunk This Movie!"
88 Rivera, Arrigucci, and Schaefer, "Intelligent Design Opponents Willing to Debate."
89 Frankowski, *Expelled.*
90 Wells, *Politically Incorrect Guide*, 129.
91 Luskin, "What Is Intelligent Design?," 167.
92 Hedin, *Canceled Science*, 186 (emphasis in original).
93 The Numbers, "Expelled: No Intelligence Allowed."
94 Axe, *Undeniable*, 26.
95 Leisola and Witt, *Heretic.*
96 Behe, foreword to *Darwin on Trial*, 12.
97 Wells, *Zombie Science*, 13, 16, 189.
98 West, *Darwin Day in America*, 241–242.
99 Schaefer, foreword to *Mere Creation*, 10.
100 Johnson, *Darwin on Trial*, 192.
101 Johnson, *Reason in the Balance*, 11.

102 Tipler, "Refereed Journals," 122.
103 Hofstadter, *Anti-Intellectualism in American Life*, 34, 130.
104 Examples might include McCarraher, *Enchantments of Mammon*; Graeber, *Utopia of Rules*; and Noble, *America by Design*.
105 Chapman, *Politicians*, 55.
106 Chapman, 146.
107 Chapman, 15.
108 Axe et al., *The Price of Panic*, 157.
109 Axe et al., 75–76.
110 Axe et al., 83.
111 Axe et al., 189–190.
112 Fuller, foreword to *Theistic Evolution*, 29.
113 Axe et al., *Price of Panic*, 187.
114 Meyer, *Darwin's Doubt*, 382–383.
115 Meyer, *Signature in the Cell*, 20.
116 Axe, *Undeniable*, 21.
117 Axe, 55.
118 Axe, 59.
119 Axe, 59–64.
120 Meyer, *Signature in the Cell*, 340.
121 Gibbs and Jenkins, *Game Plan for Life*, 79.
122 Gibbs and Jenkins, 82.
123 Gibbs and Jenkins, 93.
124 J. Evans, *Morals Not Knowledge*, 90–92.
125 Berlinski, "Deniable Darwin," 51.
126 Johnson, "Thinking Problem in HIV-Science."
127 Johnson, "Overestimating AIDS."
128 Dembski, "Tribeca Film Festival Cancels the Film *Vaxxed*."
129 Bethell, *Questioning Einstein*.
130 Axe et al., *Price of Panic*, xiii–xvii.
131 Axe et al., 14.
132 Axe et al., 202.
133 T. Nichols, *Death of Expertise*, 111.
134 T. Nichols, 54.
135 T. Nichols, 171–172, 192.
136 Wells, *Zombie Science*, 15–17.
137 Zeller, "Conservapedia."
138 Siegel, "Conservapedia."
139 "Intelligent Design," Wikipedia; "Intelligent Design," Conservapedia.
140 "Evolution and Liberalism," Conservapedia.
141 "Climate Agenda," Conservapedia.
142 Axe et al., *Price of Panic*, 40.
143 J. Evans, *Morals Not Knowledge*, 5.

144 Quoted in J. Evans, 19.
145 Meyer, *Return of the God Hypothesis*, 433
146 Shaw, "Pressure to Conform," 524.
147 Moreland, "How Theistic Evolution Kicks Christianity Out of the Plausibility Structure," 657.
148 J. Evans, *Morals Not Knowledge*, 126–127.
149 J. Evans, 147.
150 Dembski, "Launch of the Walter Bradley Center."
151 See Kurzweil, *Singularity Is Near*; and Bostrom, *Superintelligence*.
152 Dembski, "Launch of the Walter Bradley Center."
153 Walter Bradley Center for Natural and Artificial Intelligence, "Mission."
154 Dembski and Marks, *For a Greater Purpose*. See *Mind Matters* online, https://mindmatters.ai.
155 Taulli, *Artificial Intelligence Basics*, 15–16.
156 Marcus and Davis, *Rebooting AI*, 41.
157 Lee and Chen, *AI 2041*, 437.
158 Taulli, *Artificial Intelligence Basics*, 180.
159 Mitchell, *Artificial Intelligence*, 38.
160 Mitchell, 97.
161 Taulli, *Artificial Intelligence Basics*, 87.
162 Murgia, "Sci-fi Writer Ted Chiang."
163 For more on Google's use of AI for advertising and customer surveillance, see Zuboff, *Age of Surveillance Capitalism*.
164 Marks, "Will Intelligent Machines Rise Up and Overtake Humanity?," 443.
165 Dembski, *Being as Communion*, 48–49n4.
166 Marks, *Non-Computable You*, 61.
167 Marks, 37–38.
168 Erik Larson, *Myth of Artificial Intelligence*, 1–3.
169 Erik Larson, 127.
170 Erik Larson, 168–172.
171 Erik Larson, 166.
172 Smith, *AI Delusion*, 99.
173 Smith, *Distrust*, 17.
174 Smith, 2.
175 Smith, 139.
176 Erik Larson, *Myth of Artificial Intelligence*, 269–271.
177 Marks, "Will Intelligent Machines Rise Up and Overtake Humanity?," 447.
178 Dembski, "Launch of the Walter Bradley Center."
179 Marcus, "Deep Learning Is Hitting a Wall."
180 Marcus and Davis, *Rebooting AI*, 178.
181 Feser, *Philosophy of Mind*.
182 Ecklund and Scheitle, *Religion vs. Science*, 94–95, 103.
183 Humes, *Monkey Girl*, 310–311.

CONCLUSION

1 McDowell, "How Is the Intelligent Design Movement Doing?"
2 Dembski, "Official Retirement from Intelligent Design." Dembski repeatedly stressed that his retirement does not mean he has repudiated his ID works or arguments. See Dembski, "Retirement ≠ Repudiation."
3 Dembski, Thomas, and Vikander, *Dalko*.
4 Klinghoffer, "Remembering Phillip E. Johnson."
5 Reynolds, "Gandalf Has Gone to the West."
6 Luskin, "Ham-Nye Creation Debate."
7 Lovan, "Noah's Ark Project Spurred by Evolution Debate."
8 McDowell, "How Is the Intelligent Design Movement Doing?"
9 Barham, "William Dembski Interview."
10 Trollinger and Trollinger, *Righting America at the Creation Museum*, 261n39.
11 Bielo, *Ark Encounter*, 82.
12 Bielo, 176.
13 Rogan, interview with Steven Meyer.
14 Ecklund and Scheitle note (*Religion vs. Science*, 94–95) that the correlation between anti-evolution and climate denialism is not always due to religious beliefs but very often political affiliation. In the case of ID, however, the two overlap to such an extent that even if there is no logical linkage between the two, there is certainly a cultural one.
15 Geisler, *Creation and the Courts*, 296–297.
16 Kennedy v. Bremerton School District, 597 U.S. 507 (2022), slip op., 2.
17 Chesterton, *Napoleon of Notting Hill*, 5.
18 Dembski, "William Dembski: Why I'm Returning to the Front Lines."
19 Meyer, *Return of the God Hypothesis*, 436–437.
20 Meyer, 450.
21 Chiang, *Exhalation*, 259–260, 262, 265, 267, 269.
22 Chiang, 349.
23 James, "Philosophical Conceptions and Practical Effects," 299.

BIBLIOGRAPHY

Alexander, Scott. "New Atheism: The Godlessness That Failed." *Slate Star Codex* (blog), October 30, 2019. https://slatestarcodex.com.

Ali, Ayaan Hirsi. *Infidel*. New York: Atria, 2007.

Ali, Ayaan Hirsi. "Why I Am Now a Christian." *UnHerd*, November 11, 2023. https://unherd.com.

Amarasingam, Amarnath, ed. *Religion and the New Atheism: A Critical Appraisal*. Boston: Brill, 2010.

Anderson, Ross. "Evolution of a Think Tank." *Crosscut*, April 28, 2008. https://web.archive.org.

Axe, Douglas. "Extreme Functional Sensitivity to Conservative Amino Acid Changes on Enzyme Exteriors." *Journal of Molecular Biology* 301, no. 3 (2000): 585–595. https://doi.org/10.1006/jmbi.2000.3997.

Axe, Douglas. "Is Our Intuition of Design in Nature Correct?" In *The Comprehensive Guide to Science and Faith: Exploring the Ultimate Questions About Life and the Cosmos*, edited by William A. Dembski, Casey Luskin, and Joseph M. Holden. Eugene, OR: Harvest House, 2021.

Axe, Douglas. *Undeniable: How Biology Confirms Our Intuition That Life Is Designed*. New York: HarperCollins, 2016.

Axe, Douglas, William M. Briggs, and Jay W. Richards. *The Price of Panic: How the Tyranny of Experts Turned a Pandemic into a Catastrophe*. Washington, DC: Regnery, 2020.

Axe, Douglas, Brendan Dixon, and Ann Gauger. "Good Science Will Come." *New Scientist*, January 13, 2007. www.newscientist.com.

Ayala, Francisco J. "Design Without Designer: Darwin's Greatest Discovery." In *Debating Design: From Darwin to DNA*, edited by Michael Ruse and William A. Dembski. New York: Cambridge University Press, 2004.

Bailey, Ronald. "BioLogic Institute = Tobacco Institute." *Reason*, December 19, 2006. https://reason.com.

Bailey, Ronald. "Flunk This Movie!" *Reason*, July 2008. https://reason.com/.

Bailey, Ronald. "Origin of the Specious: Why Do Neoconservatives Doubt Darwin?" *Reason*, July 1997. https://reason.com.

Banerjee, Neela. "School Board Sued on Mandate for Alternative to Evolution." *New York Times*, December 20, 2004. www.nytimes.com.

Barbour, Ian. *Religion and Science: Historical and Contemporary Issues*. San Francisco: HarperCollins, 1990.

Barham, James. "William Dembski Interview." The Best Schools, January 14, 2019. https://thebestschools.org.

Bateson, William. *Materials for the Study of Variation: Treated with Especial Regard to Discontinuity in the Origin of Species*. Cambridge: Cambridge University Press, 2010.

Baylor University External Review Committee. "The External Review Committee Report: Baylor University." October 16, 2000. web.archive.org.

Beckwith, Francis J. *Law, Darwinism, and Public Education: The Establishment Clause and the Challenge of Intelligent Design*. Lanham, MD: Rowman and Littlefield, 2003.

Beckwith, Francis J. *Never Doubt Thomas: The Catholic Aquinas as Evangelical and Protestant*. Waco, TX: Baylor University Press, 2019.

Beckwith, Francis J. "Review—*The Creationists: From Scientific Creationism to Intelligent Design*, by Ronald Numbers." *Journal of Law and Religion* 23, no. 2 (2008): 735–738. https://doi.org/10.1017/S0748081400002423.

Behe, Michael J. "Bullet Points for Jerry Coyne." *Evolution News & Science Today*, March 12, 2019. https://evolutionnews.org.

Behe, Michael J. *Darwin Devolves: The New Science About DNA that Challenges Evolution*. New York: HarperCollins, 2019.

Behe, Michael J. *Darwin's Black Box: The Biochemical Challenge to Evolution*. New York: Free Press, 1996.

Behe, Michael J. *The Edge of Evolution: The Search for the Limits of Darwinism*. New York: Free Press, 2008.

Behe, Michael J. Foreword to *Darwin on Trial: 20th Anniversary Edition*, by Phillip E. Johnson. Downers Grove, IL: InterVarsity, 2010.

Behe, Michael J. "From Muttering to Mayhem: How Phillip Johnson Got Me Moving." In *Darwin's Nemesis: Phillip Johnson and the Intelligent Design Movement*, edited by William Dembski. Downers Grove, IL: InterVarsity, 2006.

Behe, Michael J. "How Does Irreducible Complexity Challenge Darwinism?" In *The Comprehensive Guide to Science and Faith: Exploring the Ultimate Questions About Life and the Cosmos*, edited by William A. Dembski, Casey Luskin, and Joseph M. Holden. Eugene, OR: Harvest House, 2021.

Behe, Michael J. *A Mousetrap for Darwin: Michael J. Behe Answers His Critics*. Seattle: Discovery Institute Press, 2020.

Behe, Michael J. "Teach Evolution—and Ask Hard Questions." *New York Times*, August 13, 1999. www.nytimes.com.

Belluck, Pam. "Board for Kansas Deletes Evolution from Curriculum." *New York Times*, August 12, 1999. www.nytimes.com.

"Berkeley's Radical: An Interview with Phillip Johnson." *Touchstone Magazine*, November 2000. https://arn.org.

Berlinski, David. "The Deniable Darwin." In *The Deniable Darwin and Other Essays*, edited by David Klinghoffer. Seattle: Discovery Institute Press, 2009.

Berlinski, David. *The Devil's Delusion: Atheism and Its Scientific Pretensions*. New York: Basic Books, 2009.

Berlinski, David. "Has Darwin Met His Match?" *Commentary*, December 2002. www.commentary.org.

Bethell, Tom. *Darwin's House of Cards: A Journalist's Odyssey Through the Darwin Debates*. Seattle: Discovery Institute Press, 2016.

Bethell, Tom. *The Politically Incorrect Guide to Science*. Washington, DC: Regnery, 2005.

Bethell, Tom. *Questioning Einstein: Is Relativity Necessary?* Pueblo West, CO: Vales Lake, 2009.

Bielo, James. *Ark Encounter: The Making of a Creationist Theme Park*. New York: New York University Press, 2018.

Biever, Celeste. "The God Lab." *New Scientist* 192, no. 2582 (2006): 8–11. https://doi.org/10.1016/S0262-4079(06)61360-2.

Biological Society of Washington. "Statement from the Council of the Biological Society of Washington." October 10, 2004. web.archive.org.

Bird, Wendell R. *The Origin of Species Revisited: The Theories of Evolution and of Abrupt Appearance*. 2 vols. Nashville, TN: Regency, 1991.

Birzer, Bradley. *Russell Kirk: American Conservative*. Lexington: University of Kentucky Press, 2015.

Bostrom, Nick. *Superintelligence: Path, Dangers, Strategies*. Oxford: Oxford University Press, 2014.

Bowler, Peter J. "Darwinism and the Argument from Design: Suggestions for a Reevaluation." *Journal of the History of Biology* 10, no. 1 (1977): 29–43. https://doi.org/10.1007/BF00126093.

Bowler, Peter J. *The Eclipse of Darwinism: Anti-Darwinian Evolutionary Theories in the Decades Around 1900*. Baltimore: Johns Hopkins University Press, 1983.

Bowler, Peter J. *Evolution: The History of an Idea*. Berkeley: University of California Press, 2009.

Bowler, Peter J. *The Non-Darwinian Revolution: Reinterpreting a Historical Myth*. Baltimore: Johns Hopkins University Press, 1988.

Bradley, Walter L., Roger L. Olsen, and Charles B. Thaxton. *The Mystery of Life's Origin: Reassessing Current Theories*. New York: Philosophical Library, 1984.

Bradley, Walter L., Roger L. Olsen, and Charles B. Thaxton. *The Mystery of Life's Origin: The Continuing Controversy*. Edited by David Klinghoffer. Seattle: Discovery Institute, 2020.

Branch, Glenn. "Farewell to the Santorum Amendment?" *Reports of the National Center for Science Education* 22, nos. 1–2 (2002). https://ncse.ngo.

Branch, Glenn. "The Latest 'Intelligent Design' Journal." *Reports of the National Center for Science Education* 30, no. 6 (2010). https://ncse.ngo.

Brooke, John Hedley. "Richard Owen, William Whewell, and the *Vestiges*." *British Journal for the History of Science* 10, no. 2 (1977): 132–145. https://doi.org/10.1017/S0007087400015387.

Brumley, Larry D. "Dembski Relieved of Duties as Polanyi Center Director." Baylor University, October 19, 2000. www.baylor.edu.

Bryan, William Jennings. "God and Evolution: Charge That American Teachers of Darwinism 'Make the Bible a Scrap of Paper.'" *New York Times*, February 26, 1922.

Bryan, William Jennings. *In His Image: James Sprunt Lectures*. 1922. Project Gutenberg, 2004. https://gutenberg.org.

Buchanan, Pat. "Darwin's Pyrrhic Victory." *Real Clear Politics*, December 28, 2005. www.realclearpolitics.com.

Buchanan, Pat. "Making a Monkey out of Darwin." *WorldNetDaily*, June 29, 2009. www.wnd.com.

Buckley, William F. *God and Man at Yale*. Washington, DC: Regnery, 1951.

Buckley, William F. "So Help Us Darwin." *National Review*, February 16, 2007. www.nationalreview.com.

Bulgakov, Sergei. *The Bride of the Lamb*. Translated by Boris Jakim. Grand Rapids, MI: Eerdmans, 2002.

Bullivant, Stephen. "The New Atheism and Sociology: Why Here? Why Now? What Next?" In *Religion and the New Atheism: A Critical Appraisal*, edited by Amarnath Amarasingam. Boston: Brill, 2010.

Bunting, Madeleine. "Why the Intelligent Design Lobby Thanks God for Richard Dawkins." *The Guardian*, March 26, 2006. www.theguardian.com.

Burress, Charles. "Teaching Evolution as Theory Not Fact: Intelligent Design Booster Speaks Out." *San Francisco Chronicle*, December 12, 2004. www.sfgate.com.

Calvin, Melvin. Foreword to *Biochemical Predestination*, edited by Dean H. Kenyon and Gary Steinman. New York: McGraw-Hill, 1969.

Came, Daniel. "Richard Dawkins's Refusal to Debate Is Cynical and Anti-Intellectualist." *The Guardian*, October 22, 2011. www.theguardian.com.

Campbell, John Angus. "Report on the Mere Christian Conference." *Origins and Design* 18, no. 1 (1997). https://arn.org.

Carnes, Tony. "Intelligent Design: Design Interference." *Christianity Today*, December 4, 2000. www.christianitytoday.com.

Caudill, Edward. *Intelligently Designed: How Creationists Built the Campaign Against Evolution*. Urbana: University of Illinois Press, 2013.

Chambers, Robert. *Vestiges of the Natural History of Creation*. Edited by James A. Secord. Chicago: University of Chicago Press, 1994.

Chapman, Bruce. "An Intelligent Discussion About Life." Discovery Institute, April 17, 2008. www.discovery.org.

Chapman, Bruce. *Politicians: The Worst Kind of People to Run the Government, Except for All the Others*. Seattle: Discovery Institute Press, 2018.

Chapman, Bruce. Postscript to *Mere Creation: Science, Faith, and Intelligent Design*, edited by William A. Dembski. Downers Grove, IL: InterVarsity, 1998.

Chapman, Matthew. *40 Days and 40 Nights: Darwin, Intelligent Design, God, OxyContin, and Other Oddities on Trial in Pennsylvania*. New York: HarperCollins, 2007.

Chesterton, G. K. *The Napoleon of Notting Hill*. New York: Dover, 1991.

Chiang, Ted. *Exhalation: Stories*. New York: Knopf, 2019.

Clayton, Philip. *Religion and Science: The Basics*. New York: Routledge, 2018.

"Climate Agenda." Conservapedia. Accessed August 2, 2020. www.conservapedia.com.

Cobbett, William. *Selections from Cobbett's Political Works: Being a Complete Abridgement of the 100 Volumes Which Comprise the Writings of "Porcupine" and the "Weekly Political Register." With Notes, Historical and Explanatory. By John M. Cobbett and James P. Cobbett*. Vol. 2. London: Ann Cobbett, 1835. https://oll.libertyfund.org.

Collins, C. John. "How to Think About God's Action in the World." In *Theistic Evolution: A Scientific, Philosophical, and Theological Critique*, edited by Anne K. Gauger, Wayne Gruden, Stephen C. Meyer, J. P. Moreland, and Christopher Shaw. Wheaton, IL: Crossway Books, 2017.

Collins, Francis. *The Language of God: A Scientist Presents Evidence for Belief*. New York: Free Press, 2006.

Colson, Charles W. Foreword to *The Design Revolution: Answering the Toughest Questions About Intelligent Design*, by William A. Dembski. Downers Grove, IL: InterVarsity, 2004.

"Conservatives, Darwin, & Design: An Exchange." *First Things*, November 2000. www.firstthings.com.

Cootsona, Gregory S. *Mere Science and Christian Faith: Bridging the Divide with Emerging Adults*. Downers Grove, IL: InterVarsity, 2018.

Cotter, Christopher R., Philip Andrew Quadrio, and Jonathan Tuckett, eds. *New Atheism: Critical Perspectives and Contemporary Debates*. Cham, Switzerland: Springer, 2017.

Couzin-Frankel, Jennifer. "Bush Backs Teaching Intelligent Design." *Science Magazine*, August 2, 2005. www.sciencemag.org.

Coyne, Jerry A. "Creationism by Stealth." *Nature* 410 (2001): 745–746. https://doi.org/10.1038/35071144.

Coyne, Jerry A. "God in the Details." *Nature* 383 (1996): 227–228. https://doi.org/10.1038/383227a0.

Coyne, Jerry A. "ID Craziness: Diarrhea and the Appendix Are Signs of Intelligent Design." *Why Evolution Is True* (blog), May 30, 2019. https://whyevolutionistrue.com.

Coyne, Jerry A. "Intelligent Design Gets Even Dumber." *Washington Post*, March 8, 2019. www.washingtonpost.com.

Coyne, Jerry A. "Jason Rosenhouse Pronounces Intelligent Design Dead." *Why Evolution Is True* (blog), November 30, 2011. https://whyevolutionistrue.com.

Coyne, Jerry A. *Why Evolution Is True*. New York: Penguin, 2010.

Cunningham, Conor. *Darwin's Pious Idea: Why the Ultra-Darwinists and Creationists Both Get It Wrong*. Grand Rapids, MI: Eerdmans, 2010.

Currid, John D. "Theistic Evolution Is Incompatible with the Teachings of the Old Testament." In *Theistic Evolution: A Scientific, Philosophical, and Theological Critique*, edited by Anne K. Gauger, Wayne Gruden, Stephen C. Meyer, J. P. Moreland, and Christopher Shaw. Wheaton, IL: Crossway Books, 2017.

Darwin, Charles. Letter to Asa Gray, May 22, 1860. *Darwin Correspondence Project*. www.darwinproject.ac.uk.

Darwin, Charles. Letter to J. D. Hooker, February 1, 1871. *Darwin Correspondence Project*. www.darwinproject.ac.uk.

Darwin, Charles. *The Origin of Species: A Facsimile of the First Edition*. Cambridge, MA: Harvard University Press, 2001.

Davis, Edward B. "Fundamentalism and Folk Science Between the Wars." *Religion and American Culture: A Journal of Interpretation* 5, no. 2 (1995): 217–248. https://doi.org/10.2307/1123857.

Dawkins, Richard. *The Blind Watchmaker: Why the Evidence of Evolution Reveals a Universe Without Design*. New York: Norton, 1996.

Dawkins, Richard. *Brief Candle in the Dark: My Life in Science*. New York: HarperCollins, 2015.

Dawkins, Richard. *The God Delusion*. New York: Mariner Books, 2006.

Dawkins, Richard. *The Greatest Show on Earth: The Evidence for Evolution*. New York: Free Press, 2010.

Dawkins, Richard. "Inferior Design." *New York Times*, July 1, 2007. www.nytimes.com.

Dawkins, Richard. "Religion's Misguided Missiles." *The Guardian*, September 15, 2001. www.theguardian.com.

Dawkins, Richard. *River Out of Eden: A Darwinian View of Life*. New York: Basic Books, 1995.

Dawkins, Richard. "Time to Stand Up." *New Humanist*, 2001. https://newhumanist.org.uk.

"Days Six & Seven: Transcript of Scopes Trial, Friday July 17 & Monday July 20, 1925." *The Scopes Trial (1925)*. Clarence Darrow Digital Collection, University of Minnesota Law Library. https://law.umn.edu.

Dembski, William A. *Being as Communion: A Metaphysics of Information*. Burlington, VT: Ashgate, 2014.

Dembski, William A. "Dealing with the Backlash Against Intelligent Design." In *Darwin's Nemesis: Phillip Johnson and the Intelligent Design Movement*, edited by William A. Dembski. Downers Grove, IL: InterVarsity, 2006.

Dembski, William A. *The Design Inference: Eliminating Chance Through Small Probabilities*. New York: Cambridge University Press, 1998.

Dembski, William A. *The Design Revolution: Answering the Toughest Questions About Intelligent Design*. Downers Grove, IL: InterVarsity, 2004.

Dembski, William A. *The End of Christianity: Finding a Good God in an Evil World*. Nashville, TN: B&H, 2009.

Dembski, William A. *Intelligent Design: The Bridge Between Science and Theology*. Downers Grove, IL: InterVarsity, 1999.

Dembski, William A. "Intelligent Design Coming Clean." Access Research Network, November 17, 2000. https://arn.org/.

Dembski, William A. Introduction to *Mere Creation: Science, Faith, and Intelligent Design*, edited by William A. Dembski. Downers Grove, IL: InterVarsity, 1998.

Dembski, William A. "Is Intelligent Design a Form of Natural Theology?" *Metanexus*, May 11, 2001. https://metanexus.net.

Dembski, William A. "Launch of the Walter Bradley Center for Natural and Artificial Intelligence." *Bill Dembski* (blog), July 12, 2018. https://billdembski.com.

Dembski, William A. "Life After Dover." *Uncommon Descent* (blog), September 30, 2005. https://uncommondescent.com.

Dembski, William A. "Life Work of Bill Dembski." *Bill Dembski* (blog). Accessed January 6, 2025. https://billdembski.com.

Dembski, William A. "Naturalism's Argument from Invincible Ignorance: A Response to Howard Van Till." Discovery Institute, September 9, 2002. www.discovery.org.

Dembski, William A. *No Free Lunch: Why Specified Complexity Cannot Be Purchased Without Intelligence*. Lanham, MD: Rowman and Littlefield, 2002.

Dembski, William A. "Official Retirement from Intelligent Design." *Bill Dembski* (blog), September 23, 2016. https://billdembski.com.

Dembski, William A. "Polanyi Center Press Release." Email. October 17, 2001. www.anti-evolution.org.

Dembski, William A. Preface to *Darwin's Nemesis: Phillip Johnson and the Intelligent Design Movement*, edited by William A. Dembski. Downers Grove, IL: InterVarsity, 2006.

Dembski, William A. "Retirement ≠ Repudiation." *Bill Dembski* (blog), June 15, 2020. https://billdembski.com.

Dembski, William A. "Thomas Reid Lectures on Natural Theology—New Edition." *Bill Dembski* (blog), June 21, 2020. https://billdembski.com.

Dembski, William A. "Tribeca Film Festival Cancels the Film *Vaxxed*." *Bill Dembski* (blog), April 1, 2016. https://billdembski.com.

Dembski, William A. "William Dembski: Why I'm Returning to the Front Lines of Intelligent Design." *Evolution News & Science Today*, February 16, 2021. www.evolutionnews.org.

Dembski, William A., Wayne J. Downs, and Fr. Justin B. A. Frederick, eds. *The Patristic Understanding of Creation: An Anthology of Writings from the Church Fathers on Creation and Design*. Nashville, TN: Erasmus, 2008.

Dembski, William A., and Robert J. Marks II. *For a Greater Purpose: The Life and Legacy of Walter Bradley*. Nashville, TN: Erasmus, 2020.

Dembski, William A., Alex Thomas, and Brian Vikander. *Dalko: The Untold Story of Baseball's Fastest Pitcher*. Nashville, TN: Influence, 2020.

Dennett, Daniel. *Breaking the Spell: Religion as a Natural Phenomenon*. New York: Penguin, 2006.

Denton, Michael J. "An Anti-Darwinian Intellectual Journey: Biological Order as an Inherent Property of Matter." In *Uncommon Dissent: Intellectuals Who Find Darwinism Unconvincing*, edited by William A. Dembski. Wilmington, DE: ISI Books, 2004.

Denton, Michael J. *Evolution: A Theory in Crisis*. Chevy Chase, MD: Adler and Adler, 1985.

Denton, Michael J. *Nature's Destiny: How the Laws of Biology Reveal Purpose in the Universe*. New York: Free Press, 2002.

Desmond, Adrian. *The Politics of Evolution: Morphology, Medicine, and Reform in Radical London*. Chicago: University of Chicago Press, 1989.

Desmond, Adrian, and James R. Moore. *Darwin: The Life of a Tormented Evolutionist*. New York: Norton, 1991.

DeWolf, David K., John G. West, Casey Luskin, and Jonathan Witt. *Traipsing into Evolution: Intelligent Design and the Kitzmiller vs. Dover Decision*. Seattle: Discovery Institute Press, 2006.

Discovery Institute. "Discovery Calls Dover Evolution Policy Misguided, Calls for Its Withdrawal." December 14, 2004. www.discovery.org.

Discovery Institute. "Key Resources for Parents and School Board Members." March 25, 2004. www.discovery.org.

Discovery Institute. "Major Grants Help Establish Center for Renewal of Science and Culture." August 10, 1996. https://web.archive.org.

Discovery Institute. "U.S. Congressional Committee Report: Intolerance and the Politicization of Science at the Smithsonian." December 15, 2006. www.discovery.org.

Dobzhansky, Theodosius. "Nothing in Biology Makes Sense Except in the Light of Evolution." *American Biology Teacher* 35, no. 3 (1973): 125–129. https://doi.org/10.2307/4444260.

Eberlin, Marcos. *Foresight: How the Chemistry of Life Reveals Planning and Purpose*. Seattle: Discovery Institute Press, 2019.

Ecklund, Elaine Howard, and Christopher P. Scheitle. "The Influence of Science Popularizers on the Public's View of Religion and Science: An Experimental Assessment." *Public Understanding of Science* 26, no. 1 (2017): 25–39. https://doi.org/10.1177/0963662515588432.

Ecklund, Elaine Howard, and Christopher P. Scheitle. *Religion vs. Science: What Religious People Really Think*. New York: Oxford University Press, 2010.

"Editorial: Putting First Things First." *First Things*, March 1990. www.firstthings.com.

"Editorial Policies." *BIO-Complexity*. Accessed February 27, 2021. https://bio-complexity.org.

Egnor, Michael. "Welcome, Jerry Coyne, to the Exciting Field of Intelligent Design Research." *Evolution News & Science Today*, March 3, 2019. www.evolutionnews.org.

Eldredge, Niles. *The Triumph of Evolution . . . and the Failure of Creationism*. New York: Holt, 2000.

Elmhirst, Sophie. "Is Richard Dawkins Destroying His Reputation?" *The Guardian*, June 9, 2015. www.theguardian.com.

Elsberry, Wesley R., and Nicholas Matzke. "The Collapse of Intelligent Design." In *Intelligent Design: William A. Dembski and Michael Ruse in Dialogue*, edited by Robert B. Stewart. Minneapolis: Fortress, 2007.

"Evangelical Scientists Refute Gravity with New 'Intelligent Falling' Theory." *The Onion*, August 17, 2005. www.theonion.com.

Evans, Gwen. "Reason or Faith? Darwin Expert Reflects." *University of Wisconsin–Madison News*, February 3, 2009. https://news.wisc.edu.

Evans, John H. *Morals Not Knowledge: Recasting the Contemporary U.S. Conflict Between Religion and Science*. Berkeley: University of California Press, 2018.

"Evolution and Liberalism." Conservapedia. Accessed August 2, 2020. https://web.archive.org.
"Expelled: No Intelligence Allowed." The Numbers, accessed June 24, 2020. https://the-numbers.com.
Fawcett, Edmund. *Conservatism: The Fight for a Tradition*. Princeton, NJ: Princeton University Press, 2020.
Feser, Edward. *Aristotle's Revenge: The Metaphysical Foundations of Physical and Biological Science*. Heusenstamm, Germany: editiones scholasticae, 2014.
Feser, Edward. *Five Proofs of the Existence of God*. San Francisco: Ignatius, 2017.
Feser, Edward. *Philosophy of Mind (A Beginner's Guide)*. Oxford, UK: Oneworld, 2006.
Feser, Edward. *Scholastic Metaphysics: A Contemporary Introduction*. Heusenstamm, Germany: editiones scholasticae, 2014.
Feyerabend, Paul. "How to Be a Good Empiricist: A Plea for Tolerance in Matters Epistemological." In *Philosophy of Science: The Central Issues*, edited by J. A. Cover, Martin Curd, and Christopher Pinock. New York: Norton, 2012.
Firinglinevideos. "A Firing Line Debate: Resolved: That the Evolutionists Should Acknowledge Creation." YouTube, February 7, 2017. www.youtube.com/watch?v=ITqiIQu-fbA.
Flew, Antony, and Roy Abraham Varghese. *There Is a God: How the World's Most Notorious Atheist Changed His Mind*. New York: HarperOne, 2007.
Forrest, Barbara. "Still Creationism After All These Years: Understanding and Counteracting Intelligent Design." *Integrative and Comparative Biology* 48, no. 2 (2008): 189–201. https://doi.org/10.1093/icb/icn032.
Forrest, Barbara, and Paul Gross. *Creationism's Trojan Horse: The Wedge of Intelligent Design*. New York: Oxford University Press, 2007.
Frair, Wayne. Preface to *Creation and the Courts: Eighty Years of Conflict in the Classroom and the Courtroom*, by Norman L. Geisler. Wheaton, IL: Crossway Books, 2007.
Frank, Thomas. *What's the Matter with Kansas? How Conservatives Won the Heart of America*. New York: Holt, 2004.
Frankowski, Nathan, dir. *Expelled: No Intelligence Allowed*. Vivendi Entertainment, Rocky Mountain Pictures, 2008. Film.
Frost, Robert. *The Poetry of Robert Frost: The Collected Poems, Complete and Unabridged*. Edited by Edward Connery Lathem. New York: Holt, 1969.
Fuller, Steve. Foreword to *Theistic Evolution: A Scientific, Philosophical, and Theological Critique*, edited by Anne K. Gauger, Wayne Gruden, Stephen C. Meyer, J. P. Moreland, and Christopher Shaw. Wheaton, IL: Crossway Books, 2017.
Futuyma, Douglas. "Evolution: The Hypothesis That Ranks as Fact." *New York Times*, October 12, 1986.
Gauger, Ann K., Wayne Gruden, Stephen C. Meyer, J. P. Moreland, and Christopher Shaw, eds. *Theistic Evolution: A Scientific, Philosophical, and Theological Critique*. Wheaton, IL: Crossway Books, 2017.
Geisler, Norman L. *Creation and the Courts: Eighty Years of Conflict in the Classroom and the Courtroom*. Wheaton, IL: Crossway Books, 2007.

Geisler, Norman L. *Origin Science: A Proposal for the Creation-Evolution Controversy*. Grand Rapids, MI: Baker Book House, 1987.

Geisler, Norman L. "A Review of Michael Behe's *The Edge of Evolution: The Search for the Limits of Darwinism*." *Norman Geisler* (blog). Accessed February 24, 2021. https://normangeisler.com.

Gibbs, Joe, and Jerry B. Jenkins. *Game Plan for Life: Your Personal Playbook for Success*. Carol Stream, IL: Tyndale House, 2009.

Gilbert, Scott F. "The Aerodynamics of Flying Carpets: Why Biologists Are Loath to 'Teach the Controversy.'" In *The Panda's Black Box: Opening Up the Intelligent Design Controversy*, edited by Nathaniel C. Comfort. Baltimore: Johns Hopkins University Press, 2007.

Gilder, George. "Evolution and Me." *National Review* 58, no. 13 (2006): 29–34.

Gillem, Nicholas W. "Evolution by Jumps: Francis Galton and William Bateson and the Mechanism of Evolutionary Change." *Genetics* 159, no. 4 (2001): 1383–1392. https://doi.org/10.1093/genetics/159.4.1383.

Gish, Duane T. *Speculations and Experiments Related to Theories on the Origin of Life: A Critique*. San Diego: Institute for Creation Research, 1972.

Godfrey-Smith, Peter. *Theory and Reality: An Introduction to the Philosophy of Science*. Chicago: University of Chicago Press, 2003.

Gonzalez, Guillermo. "What Astrobiology Teaches About the Origin of Life." In *The Mystery of Life's Origin: The Continuing Controversy*, edited by David Klinghoffer. Seattle: Discovery Institute Press, 2020.

Goodstein, Laurie. "Evolution Trial in Hands of Willing Judge." *New York Times*, December 18, 2005. www.nytimes.com.

Goodstein, Laurie. "Intelligent Design Might Be Meeting Its Maker." *New York Times*, December 4, 2005. www.nytimes.com.

Gould, Stephen Jay. "Impeaching a Self-Appointed Judge." *Scientific American* 267, no. 1 (1992): 118–121.

Gould, Stephen Jay. *The Panda's Thumb: More Reflections on Natural History*. New York: Norton, 1980.

Graeber, David. *The Utopia of Rules: On Technology, Stupidity, and the Secret Joys of Bureaucracy*. New York: Melville House, 2015.

Grant, Bruce. "Intentional Deception: Intelligent Design Creationism." *Skeptic* 11, no. 2 (2004). www.skeptic.com.

Grudem, Wayne. "Biblical and Theological Introduction: The Incompatibility of Theistic Evolution with the Biblical Account of Creation and with Important Christian Doctrines." In *Theistic Evolution: A Scientific, Philosophical, and Theological Critique*, edited by Anne K. Gauger, Wayne Gruden, Stephen C. Meyer, J. P. Moreland, and Christopher Shaw. Wheaton, IL: Crossway Books, 2017.

Grudem, Wayne. "Theistic Evolution Undermines Twelve Creation Events and Several Crucial Christian Doctrines." In *Theistic Evolution: A Scientific, Philosophical, and Theological Critique*, edited by Anne K. Gauger, Wayne Gruden, Stephen C. Meyer, J. P. Moreland, and Christopher Shaw. Wheaton, IL: Crossway Books, 2017.

Ham, Ken. "Intelligent Design Is Not Enough." Answers in Genesis, August 31, 2011. https://answersingenesis.org.

Ham, Ken. "Response from Young Earth Creationism." In *Four Views on Creation, Evolution, and Intelligent Design*, edited by J. B. Stump. Grand Rapids, MI: Zondervan, 2017.

Harris, Sam. *The End of Faith: Religion, Terror, and the Future of Reason*. New York: Norton, 2005.

Harris, Sam. *Letter to a Christian Nation*. New York: Vintage, 2006.

Harris, Sam. "Science Is in the Details." *New York Times*, July 26, 2009. www.nytimes.com.

Hart, David Bentley. "Believe It or Not." *First Things*, May 2010. www.firstthings.com.

Hart, David Bentley. *The Experience of God: Being, Consciousness, Bliss*. New Haven, CT: Yale University Press, 2013.

Haught, John F. *God After Darwin: A Theology of Evolution*. New York: Routledge, 2008.

Haught, John F. "Review: The Wedge of Truth." *Reports of the National Center for Science Education* 20, no. 6 (2008). https://ncse.ngo.

Hayek, Friedrich. *The Constitution of Liberty: The Definitive Edition*. Chicago: University of Chicago Press, 2011.

Hayek, Friedrich. *The Fatal Conceit: The Errors of Socialism*. Edited by W. W. Bartley III. Chicago: University of Chicago Press, 1988.

Hedin, Eric. *Canceled Science: What Some Atheists Don't Want You to See*. Seattle: Discovery Institute Press, 2021.

Heeren, Fred. "The Lynching of Bill Dembski." *American Spectator*, November 15, 2000. www.discovery.org.

Himmelfarb, Gertrude. *Darwin and the Darwinian Revolution*. Chicago: Ivan R. Dee, 1959.

Hitchens, Christopher. *God Is Not Great: How Religion Poisons Everything*. New York: Hachette, 2007.

Hofstadter, Richard. *Anti-Intellectualism in American Life*. New York: Vintage, 1962.

Hoppe, Richard. "Two Analyses of Meyer's 'Signature in the Cell.'" *Panda's Thumb*, April 24, 2010. www.pandasthumb.org.

Howell, Scott M. "William Cobbett and American Society in the Age of the French Revolution." PhD diss., University of California, Riverside, 2001.

Hume, David. *Dialogues and Natural History of Religion*. Oxford: Oxford University Press, 1993.

Humes, Edward. *Monkey Girl: Evolution, Education, Religion, and the Battle for America's Soul*. New York: Harper Perennial, 2007.

Hunter, James Davison. *Culture Wars: The Struggle to Define America*. New York: Basic Books, 1991.

Huxley, Julian. *Evolution: The Modern Synthesis*. Cambridge, MA: MIT Press, 2010.

Huxley, Thomas Henry. "The Origin of Species." *Westminster Review*, 1860. Project Gutenberg, 2013. https://gutenberg.org.

Ingrams, Richard. *The Life and Adventures of William Cobbett*. New York: Harper Perennial, 2006.

"Intelligent Design." Conservapedia. Accessed August 2, 2020. www.conservapedia.com.

"Intelligent Design." Wikipedia. Accessed August 2, 2020. https://en.wikipedia.org.

InterVarsity Press. "BioLogos Books on Science and Christianity." Accessed February 27, 2021. https://ivpress.com.

James, William. "Philosophical Conceptions and Practical Effects." *University Chronicle* 1, no. 4 (1898): 287–310.

John Templeton Foundation. "BioLogos Public Engagement: Inviting the Church and the World to See the Harmony Between Science and Biblical Faith." Accessed February 27, 2021. www.templeton.org.

John Templeton Foundation. "Engaging the Church and the World Through BioLogos Conferences and Events." Accessed February 27, 2021. www.templeton.org.

Johnson, Phillip E. "Afterword: How to Sink a Battleship: A Call to Separate Materialist Philosophy from Empirical Science." In *Mere Creation: Science, Faith, and Intelligent Design*, edited by William A. Dembski. Downers Grove, IL: InterVarsity, 1998.

Johnson, Phillip E. "Appendix 2: Phillip E. Johnson's Position Paper on Darwinism." In *Doubts About Darwin: An Intellectual History of Intelligent Design*, by Thomas E. Woodward. Grand Rapids, MI: Baker Books, 2003.

Johnson, Phillip E. "Book Review: *The Origin of Species Revisited: The Theories of Evolution and of Abrupt Appearance*." *Constitutional Commentary* 7, no. 2 (1990): 427–434.

Johnson, Phillip E. *Darwin on Trial: 20th Anniversary Edition*. Downers Grove, IL: InterVarsity, 2010.

Johnson, Phillip E. "Evolution as Dogma: The Establishment of Naturalism." *First Things*, October 1990. www.firstthings.com.

Johnson, Phillip E. "The Final Word." In *Darwin's Nemesis: Phillip Johnson and the Intelligent Design Movement*, edited by William A. Dembski. Downers Grove, IL: InterVarsity, 2006.

Johnson, Phillip E. "Overestimating AIDS." Access Research Network, April 6, 2005. https://arn.org.

Johnson, Phillip E. *Reason in the Balance: The Case Against Naturalism in Science, Law, and Education*. Downers Grove, IL: InterVarsity, 1995.

Johnson, Phillip E. "Response to Gould." *Origins Research* 15, no. 1 (1997). https://arn.org.

Johnson, Phillip E. "The Thinking Problem in HIV-Science." In *AIDS: Virus—or Drug Induced? Contemporary Issues in Genetics and Evolution*, edited by Peter H. Duesberg. New York: Kluwer, 1996.

Johnson, Phillip E. *The Wedge of Truth: Splitting the Foundations of Naturalism*. Downers Grove, IL: InterVarsity, 2000.

Johnson, Phillip E., and John Mark Reynolds. *Against All Gods: What's Right and Wrong with the New Atheism*. Downers Grove, IL: InterVarsity, 2010.

Kalthoff, Mark, ed. *Creation and Evolution in the Early American Scientific Affiliation*. Creationism in Twentieth-Century America 10. New York: Garland, 1995.

Kazin, Michael. *A Godly Hero: The Life of William Jennings Bryan*. New York: Knopf, 2007.

Keeley, Edmund. "The Art of Poetry XIII: George Seferis." *Paris Review* 50 (1970): 56–93.

Keller, David. "The Wedge of Truth Visits the Laboratory." In *Darwin's Nemesis: Phillip Johnson and the Intelligent Design Movement*, edited by William A. Dembski. Downers Grove, IL: InterVarsity, 2006.

Kenyon, Dean H. Foreword to *The Mystery of Life's Origin*, by Walter L. Bradley, Roger L. Olsen, and Charles B. Thaxton. New York: Philosophical Library, 1984.

Kenyon, Dean H. Foreword to *What Is Creation Science?*, by Henry M. Morris and Gary E. Parker. San Diego: Creation-Life, 1982.

Kenyon, Dean H., and Percival Davis. *Of Pandas and People: The Central Question of Biological Origins*. Dallas: Haughton, 1989.

Kenyon, Dean H., and Gary Steinman. *Biochemical Predestination*. New York: McGraw-Hill, 1969.

Kirk, Russell. *The Conservative Mind: From Burke to Eliot*. Washington, DC: Regnery, 2016.

Kirk, Russell. Foreword to *Visions of Order: The Cultural Crisis of Our Time*, by Richard M. Weaver. Baton Rouge: Louisiana State University Press, 1964.

Kitcher, Philip. *Abusing Science: The Case Against Creationism*. Cambridge, MA: MIT Press, 1983.

Klinghoffer, David. "The Branding of a Heretic." *Wall Street Journal*, January 28, 2005. www.wsj.com.

Klinghoffer, David, ed. *The Deniable Darwin and Other Essays*. Seattle: Discovery Institute Press, 2009.

Klinghoffer, David. "Introduction: Intelligent Design's Original Edition." In *The Mystery of Life's Origin: The Continuing Controversy*, edited by David Klinghoffer. Seattle: Discovery Institute Press, 2020.

Klinghoffer, David. "Is the Market for Articles That Ask 'Is Intelligent Design Dead?' Dead?" *Evolution News & Science Today*, April 15, 2016. https://evolutionnews.org.

Klinghoffer, David. "Mere Manipulation—Using C. S. Lewis to Pitch Evolution to Christians." *Evolution News & Science Today*, June 5, 2018. https://evolutionnews.org.

Klinghoffer, David. "The New Rallying Cry: 'Intelligent Design Is Dead!'" *Evolution News & Science Today*, December 2, 2011. https://evolutionnews.org.

Klinghoffer, David. "'No Real Conflict When One Side Gives Up': Richard Weaver and the Darwin Debate." *Evolution News & Science Today*, May 25, 2010. https://evolutionnews.org.

Klinghoffer, David. "Remembering Phillip E. Johnson (1940–2019): The Man Who Lit the Match." *Evolution News & Science Today*, November 3, 2019. https://evolutionnews.org.

Klinghoffer, David, ed. *Signature of Controversy: Responses to Critics of Signature in the Cell*. Seattle: Discovery Institute Press, 2011.

Koons, Robert C. "The Check Is in the Mail: Why Darwinism Fails to Inspire Confidence." In *Uncommon Dissent: Intellectuals Who Find Darwinism Unconvincing*, edited by William A. Dembski. Wilmington, DE: ISI Books, 2004.

Koons, Robert C., and Logan Paul Gage. "St. Thomas Aquinas on Intelligent Design." *Proceedings of the American Catholic Philosophical Association* 85 (2011): 79–97. https://doi.org/10.5840/acpaproc2011858.

Kristol, Irving. *Neoconservatism: The Autobiography of an Idea—Selected Essays 1949–1995*. New York: Free Press, 1995.

Kristol, Irving. "Room for Darwin and the Bible." *New York Times*, September 30, 1986.

Kuhn, Thomas S. *The Structure of Scientific Revolutions*. Chicago: University of Chicago Press, 1962.

Kurzweil, Ray. *The Singularity Is Near: When Humans Transcend Biology*. New York: Penguin, 2006.

Kushiner, James M. "The Measure of Design: A Conversation About the Past, Present, & Future of Darwinism & Design." *Touchstone*, July–August 2004. www.touchstonemag.com.

Lang, Gregory I., and Amber M. Rice. "Evolution Unscathed: *Darwin Devolves* Argues on Weak Reasoning That Unguided Evolution Is a Destructive Force, Incapable of Innovation." *Evolution: International Journal of Organic Evolution* 73, no. 4 (2019): 862–868. https://doi.org/10.1111/evo.13710.

Larson, Edward J. *Summer for the Gods: The Scopes Trial and America's Continuing Debate over Science and Religion*. New York: Basic Books, 2006.

Larson, Edward J. *Trial and Error: The American Controversy over Creation and Evolution*. New York: Oxford University Press, 2003.

Larson, Erik J. *The Myth of Artificial Intelligence: Why Computers Can't Think the Way We Do*. Cambridge, MA: Harvard University Press, 2021.

Lazcano, Antonio. "Historical Developments of Origins Research." *Cold Springs Harbor Perspectives in Biology* 2, no. 11 (2010): 1–16. https://doi.org/10.1101/cshperspect.a002089.

Lebo, Lauri. *The Devil in Dover: An Insider's Story of Dogma v. Darwin in Small-Town America*. New York: New Press, 2008.

Lee, Kai-Fu, and Chen Quifan. *AI 2041: Ten Visions for Our Future*. New York: Crown Currency, 2021.

Leisola, Matti, and Jonathan Witt. *Heretic: One Scientist's Journey from Darwin to Design*. Seattle: Discovery Institute Press, 2018.

LeMahieu, D. L. *The Mind of William Paley: A Philosopher and His Age*. Lincoln: University of Nebraska Press, 1976.

Lemonick, Michael D. "Dumping on Darwin: Pat Buchanan's Attacks on the Teaching of 'Godless Evolution' Tap Rich Vein of Unscientific Thought." *Time*, March 18, 1996. https://time.com.

Lents, Nathan. "Behe's Last Stand: The Lion of Intelligent Design Roars Again." *Skeptic*, March 6, 2019. www.skeptic.com.

Lewin, Roger. "Creationism on the Defensive in Arkansas: A High-Powered Battery of Lawyers and Scientists Challenges Arkansas' 'Creation Science' Law." *Science* 215, no. 4528 (1982): 33–34. https://doi.org/10.1126/science.215.4528.33.

Lewontin, Richard. "Billions and Billions of Demons." *New York Review of Books*, January 9, 1997. www.nybooks.com.

Linker, Damon. *The Theocons: Secular America Under Siege*. New York: Anchor Books, 2007.

Livingstone, David N., and Mark A. Noll. "B. B. Warfield (1851–1921): A Biblical Inerrantist as Evolutionist." *Isis* 91, no. 2 (2000): 283–304. https://doi.org/10.1086/384722.

Lovan, Dylan. "Noah's Ark Project Spurred by Evolution Debate." *Associated Press*, February 26, 2014. https://kcby.com.

Lundmark, Evelina, and Stephen LeDrew. "Unorganized Atheism and the Secular Movement: Reddit as Site for Studying 'Lived Atheism.'" *Social Compass* 66, no. 1 (2019): 112–129. https://doi.org/10.1177/0037768618816096.

Luskin, Casey. "The Ham-Nye Creation Debate: A Huge Missed Opportunity." *EvolutionNews & Science Today*, February 4, 2014. https://evolutionnews.org.

Luskin, Casey. "How Bright Is the Future of Intelligent Design?" *Evolution News & Science Today*, December 23, 2011. https://evolutionnews.org.

Luskin, Casey. "It's Time for Some Folks to Get over Dover." *Evolution News & Science Today*, December 27, 2011. https://evolutionnews.org.

Luskin, Casey. "Jonathan Wells Talks Scientific Revolutions and Counter-Revolutions." *ID the Future* (podcast), February 10, 2021 (repost). https://idthefuture.com.

Luskin, Casey. "No, Intelligent Design Doesn't Reason by Analogy; Here's Why." *Evolution News*, August 8, 2023. https://evolutionnews.org.

Luskin, Casey. "What Is Intelligent Design and How Should We Defend It?" In *The Comprehensive Guide to Science and Faith: Exploring the Ultimate Questions About Life and the Cosmos*, edited by William A. Dembski, Casey Luskin, and Joseph M. Holden. Eugene, OR: Harvest House, 2021.

Luskin, Casey. "Why Phillip Johnson Matters: A Biography." *Evolution News & Science Today*, November 17, 2011. https://evolutionnews.org.

Manning, Russell Re. Introduction to *Oxford Handbook of Natural Theology*, edited by Russell Re Manning. Oxford: Oxford University Press, 2013.

Marcus, Gary. "Deep Learning Is Hitting a Wall." *Nautilus*, March 10, 2022. https://nautil.us.

Marcus, Gary, and Ernest Davis. *Rebooting AI: Building Artificial Intelligence We Can Trust*. New York: Vintage, 2019.

Marks, Robert J., II. *Non-Computable You: What You Do That Artificial Intelligence Never Will* Seattle: Discovery Institute, 2022.

Marks, Robert J., II. "Will Intelligent Machines Rise Up and Overtake Humanity?" In *The Comprehensive Guide to Science and Faith: Exploring the Ultimate Questions About Life and the Cosmos*, edited by William A. Dembski, Casey Luskin, and Joseph M. Holden. Eugene, OR: Harvest House, 2021.

Marks, Robert J., II, and John G. West. Foreword to *The Mystery of Life's Origin: The Continuing Controversy*, edited by David Klinghoffer. Seattle: Discovery Institute Press, 2020.

Marsden, George. *Fundamentalism and American Culture*. New ed. Oxford: Oxford University Press, 2006.

Martin, Blair. "Professors Debate Legitimacy of Polanyi." *Baylor Lariat*, 1999. web.archive.org.

Matzke, Nicholas J. "Design on Trial in Dover, Pennsylvania." *Reports of the National Center for Science Education* 24, no. 5 (2004). https://ncse.ngo.

McAnulla, Stuart, Steven Kettell, and Marcus Schulzke. *The Politics of New Atheism*. New York: Routledge, 2019.

McCarraher, Eugene. *The Enchantments of Mammon: How Capitalism Became the Religion of Modernity*. Cambridge, MA: Harvard University Press, 2019.

McDowell, Sean. "How Is the Intelligent Design Movement Doing? Interview with William Dembski." *Sean McDowell* (blog), September 8, 2016. https://seanmcdowell.org.

Menuge, Angus. "Who's Afraid of ID? A Survey of the Intelligent Design Movement." In *Debating Design: From Darwin to DNA*, edited by William A. Dembski and Michael Ruse. New York: Cambridge University Press, 2004.

Meyer, Stephen C. "Danger: Indoctrination—A Scopes Trial for the 90s." *Wall Street Journal*, December 6, 1993.

Meyer, Stephen C. *Darwin's Doubt: The Explosive Origin of Animal Life and the Case for Intelligent Design*. New York: HarperOne, 2013.

Meyer, Stephen C. "The Difference It Doesn't Make: Why the 'Front-End Loaded' Concept of Design Fails to Explain the Origin of Biological Information." In *Theistic Evolution: A Scientific, Philosophical, and Theological Critique*, edited by Anne K. Gauger, Wayne Gruden, Stephen C. Meyer, J. P. Moreland, and Christopher Shaw. Wheaton, IL: Crossway Books, 2017.

Meyer, Stephen C. "Evidence of Intelligent Design in the Origin of Life." In *The Mystery of Life's Origin: The Continuing Controversy*, edited by David Klinghoffer. Seattle: Discovery Institute Press, 2020.

Meyer, Stephen C. "Letter from the Ad Hoc Origins Committee." In *Doubts About Darwin: A History of Intelligent Design*, by Thomas E. Woodward. Grand Rapids, MI: Baker Books, 2003.

Meyer, Stephen C. "The Origin of Biological Information and the Higher Taxonomic Changes." *Proceedings of the Biological Society of Washington* 117, no. 2 (2004): 213–239.

Meyer, Stephen C. "The Return of the God Hypothesis." *Journal of Interdisciplinary History* 11 (1999): 1–38.

Meyer, Stephen C. *The Return of the God Hypothesis: Three Scientific Discoveries That Reveal the Mind Behind the Universe*. New York: HarperOne, 2021.

Meyer, Stephen C. "Scientific and Philosophical Introduction: Defining Theistic Evolution." In *Theistic Evolution: A Scientific, Philosophical, and Theological Critique*, edited by Anne K. Gauger, Wayne Gruden, Stephen C. Meyer, J. P. Moreland, and Christopher Shaw. Wheaton, IL: Crossway Books, 2017.

Meyer, Stephen C. *Signature in the Cell: DNA and the Evidence for Intelligent Design*. New York: HarperOne, 2009.

Meyer, Stephen C. "Your Witness, Mr. Johnson: A Retrospective Review of *Darwin on Trial*." In *Darwin's Nemesis: Phillip Johnson and the Intelligent Design Movement*, edited by William A. Dembski. Downers Grove, IL: InterVarsity, 2006.

Meyer, Stephen C., and Paul A. Nelson. "Should Theistic Evolution Depend on Methodological Naturalism?" In *Theistic Evolution: A Scientific, Philosophical, and Theological Critique*, edited by Anne K. Gauger, Wayne Gruden, Stephen C. Meyer, J. P. Moreland, and Christopher Shaw. Wheaton, IL: Crossway Books, 2017.

Miller, Brian. "Thermodynamic Challenges to the Origin of Life." In *The Mystery of Life's Origin*, edited by David Klinghoffer. Seattle: Discovery Institute Press, 2020.

Miller, Kenneth R. "*Darwin's Black Box* Reviewed by Kenneth R. Miller." *Creation/Evolution* 16 (1996): 36–40.

Miller, Kenneth R. *Finding Darwin's God: A Scientist's Search for Common Ground Between God and Evolution*. New York: HarperCollins, 2000.

Miller, Stanley L. "A Production of Amino Acids Under Possible Primitive Earth Conditions." *Science* 117, no. 3046 (1953): 528–529. https://doi.org/10.1126/science.117.3046.528.

Mitchell, Melanie. *Artificial Intelligence: A Guide for Thinking Humans*. New York: Picador, 2019.

Moreland, J. P. "How Theistic Evolution Kicks Christianity Out of the Plausibility Structure and Robs Christians of Confidence That the Bible Is a Source of Knowledge." In *Theistic Evolution: A Scientific, Philosophical, and Theological Critique*, edited by Anne K. Gauger, Wayne Gruden, Stephen C. Meyer, J. P. Moreland, and Christopher Shaw. Wheaton, IL: Crossway Books, 2017.

Morris, Henry M. *A History of Modern Creationism*. San Diego: Master Book, 1984.

Morris, Henry M. "Intelligent Design and/or Scientific Creationism." Institute for Creation Research, April 1, 2006. www.icr.org.

Morris, Henry M. *The Long War Against God: The History and Impact of the Creation/Evolution Conflict*. Green Forest, AK: Master Books, 2000.

Morris, Henry M. "Neocreationism." Institute for Creation Research, February 1, 1998. www.icr.org.

Morris, Henry M. *Scientific Creationism*. San Diego: Creation-Life, 1974.

Morris, Henry M., and Gary E. Parker. *What Is Creation Science?* San Diego: Creation-Life, 1982.

Muller, H. J. "One Hundred Years Without Darwinism Are Enough." *School Science and Mathematics* 59, no. 4 (1959): 304–316.

Mumford, Lewis. *The Pentagon of Power: The Myth of the Machine*. Vol. 2. New York: Harcourt Brace Jovanovich, 1970.

Murgia, Madhumita. "Sci-fi Writer Ted Chiang: 'The Machines We Have Now Are Not Conscious.'" *Financial Times*, June 3, 2023.

Myers, P. Z. "The Train Wreck That Was the New Atheism." *Pharyngula* (blog), January 25, 2019. https://freethoughtblogs.com.

Nagel, Thomas. *Mind and Cosmos: Why the Materialist Neo-Darwinian Conception of Nature Is Almost Certainly False*. New York: Oxford University Press, 2012.

Nash, George H. *The Conservative Intellectual Movement in America Since 1945*. Wilmington, DE: Intercollegiate Studies Institute, 2006.

National Center for Science Education. "The Wedge Document." October 14, 2008. https://ncse.ngo.

Nelson, Paul A. "Life in the Big Tent." *Christian Research Journal* 24, no. 4 (2002). www.equip.org.

Nelson, Paul A., and John Mark Reynolds. "Young Earth Creationism." In *Three Views on Creation and Evolution*, edited by J. P. Moreland and John Mark Reynolds. Grand Rapids, MI: Zondervan, 1999.

Nettles, Tom. "Book Review—*The End of Christianity: Finding a Good God in an Evil World*." *Southern Baptist Journal of Theology* 13, no. 4 (2009): 80–85.

Neuhaus, Richard John. "Bashing Darwin, Becoming Catholic." *First Things*, March 30, 2007. www.firstthings.com.

Nichols, Alex. "New Atheism's Idiot Heirs: An Irritating Rhetoric Meets the Dumbest Possible Ideology." *The Baffler*, October 17, 2017. https://thebaffler.com.

Nichols, Tom. *The Death of Expertise: The Campaign Against Established Knowledge and Why It Matters*. New York: Oxford University Press, 2017.

Nisbet, Robert. *Prejudices: A Philosophical Dictionary*. Cambridge, MA: Harvard University Press, 1982.

Noble, David F. *America by Design: Science, Technology, and the Rise of Corporate Capitalism*. New York: Knopf, 1977.

Numbers, Ronald L. *Creation by Natural Law: Laplace's Nebular Hypothesis in American Thought*. Seattle: University of Washington Press, 1977.

Numbers, Ronald L. *The Creationists: From Scientific Creationism to Intelligent Design, Expanded Edition*. Cambridge, MA: Harvard University Press, 2006.

Numbers, Ronald L. *Darwinism Comes to America*. Cambridge, MA: Harvard University Press, 1998.

Numbers, Ronald L., and Jeff Hardin. "New Atheists." In *The Warfare Between Science and Religion: The Idea That Wouldn't Die*, edited by Jeff Hardin, Ronald L. Numbers, and Ronald A. Binzley. Baltimore: Johns Hopkins University Press, 2018.

Oakes, Edward T. "The Wedge of Truth: Splitting the Foundations of Naturalism." *First Things*, January 2001. www.firstthings.com.

Oppenheimer, Mark. "The Turning of an Atheist." *New York Times*, November 4, 2007. www.nytimes.com.

Padian, Kevin. "Lessons from the 'Intelligent Design' Trial: Explaining Evolution and Climate Science in a 'Post-Evidentiary World.'" *Integrative and Comparative Biology* 59, no. 1 (2019): 89–100. https://doi.org/10.1093/icb/icz038.

Paley, William. *Natural Theology*. New York: Oxford University Press, 2008.

Pearcey, Nancey. "Up from Materialism: An Interview with Dean Kenyon." *Bible-Science Newsletter* 27, no. 9 (1989): 6–9.

Pennock, Robert T. "DNA by Design? Stephen Meyer and the Return of the God Hypothesis." In *Debating Design: From Darwin to DNA*, edited by William A. Dembski and Michael Ruse. New York: Cambridge University Press, 2004.

Pennock, Robert T. *The Tower of Babel: The Evidence Against the New Creationism*. Cambridge, MA: MIT Press, 2000.

Pigliucci, Massimo. *How to Be a Stoic: Using Ancient Philosophy to Live a Modern Life*. New York: Basic Books, 2017.

Pigliucci, Massimo, and Gerd B. Muller, eds. *Evolution: The Extended Synthesis*. Cambridge, MA: MIT Press, 2010.

Plantinga, Alvin. "The Dawkins Confusion." *Books & Culture*, 2007. www.booksandculture.com.

Plantinga, Alvin. *Warranted Christian Belief*. New York: Oxford University Press, 2000.

Poole, Steven. "The Four Horsemen Review—Whatever Happened to 'New Atheism'?" *The Guardian*, January 31, 2019. www.theguardian.com.

Powell, Michael. "Judge Rules Against 'Intelligent Design.'" *Washington Post*, December 21, 2005. www.washingtonpost.com.

Rana, Fazale. "What Is the Biblical and Scientific Case for a Historical Adam and Eve?" In *The Comprehensive Guide to Science and Faith: Exploring the Ultimate Questions About Life and the Cosmos*, edited by William A. Dembski, Casey Luskin, and Joseph M. Holden. Eugene, OR: Harvest House, 2021.

Randolph, John L. "Myth 23—That the Soviet Launch of Sputnik Caused the Revamping of American Science Education." In *Newton's Apple and Other Myths About Science*, edited by Ronald Numbers and Kostas Kampourakis. Cambridge, MA: Harvard University Press, 2015.

Ratliffe, Evan. "The Crusade Against Evolution." *Wired*, October 1, 2004. www.wired.com.

Ratzsch, Del. "How Not to Critique Intelligent Design Theory: A Review of Niall Shanks *God, The Devil, and Darwin*." In *Darwin Strikes Back: Defending the Science of Intelligent Design*, by Thomas E. Woodward. Grand Rapids, MI: Baker Books, 2006.

Reeves, Colin R. "Bringing Home the Bacon: The Interaction of Science and Scripture Today." In *Theistic Evolution: A Scientific, Philosophical, and Theological Critique*, edited by Anne K. Gauger, Wayne Gruden, Stephen C. Meyer, J. P. Moreland, and Christopher Shaw. Wheaton, IL: Crossway Books, 2017.

Reid, Thomas. *Lectures in Natural Theology*. Nashville, TN: Erasmus, 2020.

Reynolds, John Mark. "Gandalf Has Gone to the West (Phillip E. Johnson)." *Eidos with John Mark Reynolds* (blog), November 3, 2019. www.patheos.com/blogs/eidos.

Reynolds, John Mark. "Introduction: A Mythic Life." In *Darwin's Nemesis: Phillip Johnson and the Intelligent Design Movement*, edited by William A. Dembski. Downers Grove, IL: InterVarsity, 2006.

Richards, Jay W. "*BIO-Complexity*: A New, Peer-Reviewed Science Journal, Open to the ID Debate." *Evolution News & Science Today*, May 1, 2010. https://evolutionnews.org.

Rios, Christopher M. *After the Monkey Trial: Evangelical Scientists and a New Creationism*. New York: Fordham University Press, 2014.

Rivera, Marcos, Virginia Arrigucci, and Emily Schaefer. "Intelligent Design Opponents Willing to Debate." *Iowa State Daily*, December 12, 2005. https://iowastatedaily.com.

Roberts, Michael. "Sedimentary Muddle: Review of Robert Pennock's 'The Tower of Babel.'" *Metanexus*, October 19, 2000. https://metanexus.net.

Rogan, Joe. Interview with Steven Meyer. *The Joe Rogan Experience* (podcast), July 13, 2023. www.youtube.com/watch?v=tb1Ubw1Iu5w&list=PL7Wwl5TzliiERIqbsibFY6rKdGhqQWU6U&index=1.

Rosenhouse, Jason. *Among the Creationists: Dispatches from the Anti-Evolutionist Front Line*. New York: Oxford University Press, 2012.

Rosenhouse, Jason. "Templeton Foundation and Santorum Losing Faith in ID?" *Panda's Thumb* (blog), November 14, 2005. https://pandasthumb.org.

Rosenhouse, Jason. "Twenty Years After Darwin on Trial, ID Is Dead." *Science Blogs* (blog), November 29, 2011. https://scienceblogs.com.

Rupke, Nicolaas. "Myth 13—That Darwinian Natural Selection Has Been 'The Only Game in Town.'" In *Newton's Apple and Other Myths About Science*, edited by Ronald Numbers and Kostas Kampourakis. Cambridge, MA: Harvard University Press, 2015.

Ruse, Michael. "The Argument from Design." In *The Panda's Black Box: Opening Up the Intelligent Design Controversy*, edited by Nathaniel C. Comfort. Baltimore: Johns Hopkins University Press, 2007.

Ruse, Michael. *But Is It Science? The Philosophical Question in the Creation/Evolution Controversy*. Buffalo, NY: Prometheus Books, 1988.

Sample, Ian. "Stephen Hawking Launches $100m Search for Alien Life Beyond Solar System." *The Guardian*, July 20, 2015. www.theguardian.com.

Saunders, Nicholas. *Divine Action and Modern Science*. Cambridge: Cambridge University Press, 2002.

Schaefer, Henry F., III. Foreword to *Mere Creation: Science, Faith, and Intelligent Design*, edited by William A. Dembski. Downers Grove, IL: InterVarsity, 1998.

Schmeltekopf, Donald D. *Baylor at the Crossroads: Memoirs of a Provost*. Eugene, OR: Cascade Books, 2015.

Schönborn, Christoph. "Finding Design in Nature." *New York Times*, July 7, 2005. www.nytimes.com.

"School District Challenges Darwin's Theory." *New York Times*, November 21, 2004. www.nytimes.com.

Scott, Eugenie. "Darwin Prosecuted: Review of Johnson's *Darwin on Trial*." *Creation/Evolution Journal* 13, no. 2 (1993): 36–47.

Scruton, Roger. *Conservatism: An Invitation to the Great Tradition*. New York: St. Martin's, 2018.

Secord, James A. *Victorian Sensation: The Extraordinary Publication, Reception and Secret Authorship of Vestiges of the Natural History of Creation*. Chicago: University of Chicago Press, 2000.

Sedgwick, Adam. Letter to Charles Lyell, April 9, 1845. In *The Life and Letters of Adam Sedgwick*, vol. 2, edited by John Willis Clark. Cambridge: Cambridge University Press, 2010.

Sedgwick, Adam. "*Vestiges of the Natural History of Creation*." *Edinburgh Review*, no. 165 (1845): 1–85.

Shallit, Jeffrey. "Groundless Annual Ritual of ID Self-Congratulation." *Panda's Thumb* (blog), December 19, 2014. https://pandasthumb.org.

Shallit, Jeffrey. "The Sterility of Intelligent Design." *Panda's Thumb* (blog), December 11, 2012. https://pandasthumb.org.

Shapiro, Adam. "Myth 8—That William Paley Raised Questions About Biological Origins That Were Eventually Answered by Charles Darwin." In *Newton's Apple and Other Myths About Science*, edited by Ronald Numbers and Kostas Kampourakis. Cambridge, MA: Harvard University Press, 2015.

Shaw, Christopher. "Pressure to Conform Leads to Bias in Science." In *Theistic Evolution: A Scientific, Philosophical, and Theological Critique*, edited by Anne K. Gauger, Wayne Gruden, Stephen C. Meyer, J. P. Moreland, and Christopher Shaw. Wheaton, IL: Crossway Books, 2017.

Siegel, Robert. "Conservapedia: Data for Birds of a Political Feather?" *All Things Considered*, NPR, March 13, 2007. www.npr.org.

Simonyi, Charles. "Manifesto for the Simonyi Professorship." Oxford University, May 15, 1995. www.simonyi.ox.ac.uk.

Slack, Gordy. *The Battle over the Meaning of Everything*. San Francisco: Wiley, 2007.

Smith, Gary. *The AI Delusion*. Oxford: Oxford University Press, 2018.

Smith, Gary. *Distrust: Big Data, Data Torturing, and the Assault on Science*. Oxford: Oxford University Press, 2023.

Sonleitner, Frank J. "What Did Karl Popper Really Say About Evolution?" *Creation/Evolution Journal* 6, no. 2 (1986). https://ncse.ngo.

Stenger, Victor J. *God—The Failed Hypothesis: How Science Shows That God Does Not Exist*. New York: Prometheus Books, 2008.

Stenger, Victor J. "Intelligent Design: The New Stealth Creationism." *Infidels*, 2001. https://infidels.org.

Stenger, Victor J. *The New Atheism: Taking a Stand for Science and Reason*. Amherst, NY: Prometheus Books, 2009.

Sternberg, Richard. "Procedures for the Publication of the Meyer Paper." *Richard Sternberg* (blog), October 21, 2004. web.archive.org.

Stoppard, Tom. *Rosencrantz and Guildenstern Are Dead*. New York: Grove, 1967.

Suarez-Diaz, Edna. "Molecular Evolution in Historical Perspective." *Journal of Molecular Evolution* 83 (2016): 204–213. https://doi.org/10.1007/s00239-016-9772-6.

Sullivan, Amy. "Helping Christians Reconcile God with Science." *Time*, May 2, 2009. https://time.com.

Talbot, Margaret. "Darwin in the Dock." *New Yorker*, December 5, 2005. www.newyorker.com.

Taulli, Tom. *Artificial Intelligence Basics: A Non-Technical Introduction*. New York: Apress, 2019.

Thaxton, Charles B. "Appendix 2: 1997 Update." In *The Mystery of Life's Origin: The Continuing Controversy*, edited by David Klinghoffer. Seattle: Discovery Institute Press, 2020.

Tipler, Frank J. "Refereed Journals—Do They Insure Quality of Enforce Orthodoxy?" In *Uncommon Dissent: Intellectuals Who Find Darwinism Unconvincing*, edited by William A. Dembski. Wilmington, DE: ISI Books, 2004.

"TLS Letters 09/12/09." *Times Literary Supplement*, December 9, 2009. web.archive.org.

Topham, Jonathan. "Biology in the Service of Natural Theology: Paley, Darwin, and the *Bridgewater Treatises*." In *Biology and Ideology*, edited by Denis Alexander and Ronald L. Numbers. Chicago: University of Chicago Press, 2010.

Torley, Vincent. "*Undeniable* Packs a Strong Punch, but Doesn't Land a Knockout." *VJ Torley* (blog). Accessed June 6, 2020. www.angelfire.com.

Toumey, Christopher P. *God's Own Scientists: Creationists in a Secular World*. New Brunswick, NJ: Rutgers University Press, 1994.

Tour, James M. "We're Still Clueless About the Origin of Life." In *The Mystery of Life's Origin: The Continuing Controversy*, edited by David Klinghoffer. Seattle: Discovery Institute Press, 2020.

Trollinger, Susan, and William Trollinger. *Righting America at the Creation Museum*. Baltimore: Johns Hopkins University Press, 2016.

Twain, Mark. *Roughing It*. Orinda, CA: SeaWolf, 2018.

Ury, Thane E. "Reports: Mere Creation Conference." *CEN Technical Journal* 11, no. 1 (1997): 25–30.

Vaïsse, Justin. *Neoconservatism: The Biography of a Movement*. Translated by Arthur Goldhammer. Cambridge, MA: Harvard University Press, 2010.

Van Biema, David. "God vs. Science." *Time*, November 5, 2006. https://time.com.

Van Till, Howard J. "SOBIG: A Symposium on Belief in God." *Theology and Science* 14, no. 1 (2016): 6–35. https://doi.org/10.1080/14746700.2015.1122323.

Venema, Dennis. "Intuitions, Proteins, and Evangelicals: A Response to *Undeniable*." *Sapentia*, February 5, 2018. https://henrycenter.tiu.edu.

Voegelin, Eric. *The New Order and Last Orientation*. Vol. 7 of *History of Political Ideas*. Collected Works of Eric Voegelin 25. Columbia: University of Missouri Press, 2004.

Wald, David. "Intelligent Design Meets Congressional Designers." *Skeptic* 8, no. 2 (2000): 12.

Waldrop, M. Mitchell. *The Dream Machine: J. C. R. Licklider and the Revolution That Made Computing Personal*. San Francisco: Stripe, 2018.

"Wall Street Journal Wrong on Templeton Foundation and ID." *ScienceBlogs* (blog), November 15, 2005. https://scienceblogs.com.

Wallace, Alfred Russel. *Darwinism: An Exposition of the Theory of Natural Selection with Some of Its Applications*. London: Macmillan, 1889. www.gutenberg.org.

Wallace, Paul. "Intelligent Design Is Dead: A Christian Perspective." *Huffington Post*, January 2, 2012. www.huffpost.com.

Walter Bradley Center for Natural and Artificial Intelligence. "Mission." Accessed January 6, 2025. https://bradley.center.

Ward, Keith. *The Big Questions in Science and Religion*. West Conshohocken, PA: Templeton Foundation Press, 2008.

Watson, James D. *The Double Helix: A Personal Account of the Structure of DNA*. New York: Touchstone, 2001.

Weatherly, Jack. "Creationists Lose in Arkansas: Missing Witnesses and a Divided Defense Muddled the Issue." *Christianity Today*, January 22, 1982, 28–29.

Weaver, Richard M. *Ideas Have Consequences*. Chicago: University of Chicago Press, 1948.

Weaver, Richard M. *Visions of Order: The Cultural Crisis of Our Time*. Baton Rouge: Louisiana State University Press, 1964.

Webb, David K. "But Will He Check That Recycle Bin Again?" *Origins & Design* 17, no. 2 (1996). https://arn.org.

Weinberg, Carl R. "'Ye Shall Know Them by Their Fruits': Evolution, Eschatology, and the Anticommunist Politics of George McCready Price." *Church History* 83, no. 3 (2014): 684–722. https://doi.org/10.1017/S0009640714000602.

Wells, Jonathan. *Icons of Evolution: Science or Myth? Why Much of What We Teach About Evolution Is Wrong*. Washington, DC: Regnery, 2000.

Wells, Jonathan. "Is Darwinism a Theory in Crisis?" In *The Comprehensive Guide to Science and Faith: Exploring the Ultimate Questions About Life and the Cosmos*, edited by William A. Dembski, Casey Luskin, and Joseph M. Holden. Eugene, OR: Harvest House, 2021.

Wells, Jonathan. *The Politically Incorrect Guide to Darwinism and Intelligent Design*. Washington, DC: Regnery, 2006.

Wells, Jonathan. "Textbooks Still Misrepresent the Origin of Life." In *The Mystery of Life's Origin: The Continuing Controversy*, edited by David Klinghoffer. Seattle: Discovery Institute Press, 2020.

Wells, Jonathan. "Unseating Naturalism." In *Mere Creation: Science, Faith, and Intelligent Design*, edited by William A. Dembski. Downers Grove, IL: InterVarsity, 1998.

Wells, Jonathan. *Zombie Science: More Icons of Evolution*. Seattle: Discovery Institute Press, 2017.

West, John G. *Darwin Day in America: How Our Politics and Culture Have Been Dehumanized in the Name of Science*. Wilmington, DE: ISI Books, 2015.

West, John G. *Darwinian Conservatives: The Misguided Quest*. Seattle: Discovery Institute Press, 2006.

West, John G. "Signature in the Cell Named One of Top Books of the Year by Times Literary Supplement." *Evolution News & Science Today*, November 25, 2009. https://evolutionnews.org.

Whitcomb, John C. "Review Article: Creation Science and the Physical Universe." *Grace Theological Journal* 4, no. 2 (1983): 289–296.

Whitcomb, John C., and Henry M. Morris. *The Genesis Flood: The Biblical Record and Its Scientific Implications*. 50th anniversary ed. Phillipsburg, NJ: Presbyterian and Reformed Publishing, 2011.

Wilder-Smith, A. E. *The Creation of Life: A Cybernetic Approach to Evolution*. San Diego: Master Books, 1970.

Wilder-Smith, Beate. Preface to *Let Us Reason: Insights on Creation, Suffering, and Evil*, by A. E. Wilder-Smith. Costa Mesa, CA: Word for Today, 2007.

Will, George F. *The Conservative Sensibility*. New York: Hachette Books, 2019.

Winfield, Nicole. "Vatican-Backed Conference Snubs Creationism." *NBC News*, March 5, 2009. www.nbcnews.com.

Witt, Jonathan. "Biola Honors Flew, Puts Intelligent Design in the Hot Seat." *Evolution News & Science Today*, May 11, 2006. https://evolutionnews.org.

Witt, Jonathan. "A March for Science? More a March for Meek Conformity." *The Stream*, April 13, 2018. https://stream.org.

Wolf, Gary. "The Church of the Non-Believers." *Wired*, November 1, 2006. www.wired.com.

Woodward, Thomas E. *Darwin Strikes Back: Defending the Science of Intelligent Design*. Grand Rapids, MI: Baker Books, 2006.

Woodward, Thomas E. *Doubts About Darwin: A History of Intelligent Design*. Grand Rapids, MI: Baker Books, 2003.

Wysong, Randy L. *The Creation-Evolution Controversy (Implications, Methodology and Survey of Evidence): Toward a Rational Solution*. East Lansing, MI: Inquiry, 1976.

Zeller, Shawn. "Conservapedia: See Under 'Right.'" *New York Times*, March 5, 2007. www.nytimes.com.

Zimmerman, Michael. "Creationism in New Hampshire: Attacking Science and Undermining Religious Freedom." *Huffington Post*, January 8, 2012. www.huffpost.com.

Zuboff, Shoshanna. *The Age of Surveillance Capitalism: The Fight for a Human Future at the New Frontier of Power*. New York: PublicAffairs, 2019.

INDEX

Page numbers in *italics* indicate figures.

ABOUT THE AUTHOR

C.W. HOWELL holds a PhD from Duke University in religion and has taught at Duke and Elon Universities. He writes regularly on religion, science, and technology on his website www.cwhowell.com.